Cosmology - The Past, Present and Future of the Universe

Edited by Herman J. Mosquera Cuesta and Kim Ho Yeap

Published in London, United Kingdom

Cosmology - The Past, Present and Future of the Universe
http://dx.doi.org/10.5772/intechopen.1000535
Edited by Herman J. Mosquera Cuesta and Kim Ho Yeap

Assistant to the Editors: Teoh Hui Chieh

Contributors
Avery M. Minion, David Hoxley, Herwig Dejonghe, Hui Chieh Teoh, John C. Botke, Jonathan H. Jiang, Kim Ho Yeap, Michael G. Albrow, Philip Broadbridge, Sarah Becirevic, Sho Fujita, Stuart F. Taylor

First published in London, United Kingdom, 2024 by IntechOpen
IntechOpen is the global imprint of INTECHOPEN LIMITED, registered in England and Wales, registration number: 11086078, 167-169 Great Portland Street, London, W1W 5PF, United Kingdom

British Library Cataloguing-in-Publication Data
A catalogue record for this book is available from the British Library

Additional hard and PDF copies can be obtained from orders@intechopen.com

Cosmology - The Past, Present and Future of the Universe
Edited by Herman J. Mosquera Cuesta and Kim Ho Yeap
p. cm.
Print ISBN 978-1-83769-473-0
Online ISBN 978-1-83769-472-3
eBook (PDF) ISBN 978-1-83769-474-7

For EU product safety concerns:
IN TECH d.o.o., Prolaz Marije Krucifikse Kozulić 3, 51000 Rijeka, Croatia,
info@intechopen.com or visit our website at intechopen.com.

Meet the editors

Prof. Dr. Herman J. Mosquera Cuesta is an astrophysicist from the National Institute for Space Research (INPE), a former researcher/professor at the International Center of Relativistic Astrophysics and Brazilian Center for Physics Research (ICRA-BR/CBPF), and a visiting researcher at the International Center for Relativistic Astrophysics Network (ICRANet). From 2010 to 2013, he was a professor at Acaraú Valley State University (UVA), Brazil, and the scientific coordinator of the Sobral Astro Project. He has been a visiting scientist for Colombia's National Space Program since 2015. He is currently a graduate student advisor at Valencia International University (VIU), Spain. Dr. Cuesta is a member of the Brazilian Astronomical Society, Brazilian Physical Society, and the International Society on General Relativity and Gravitation.

Kim Ho Yeap is an associate professor at Universiti Tunku Abdul Rahman, Malaysia. He is a senior member of the Institute of Electrical and Electronics Engineers (IEEE), a professional engineer registered with the Board of Engineers, Malaysia, a chartered engineer registered with the UK Engineering Council, and an Association of Southeast Asian Nations (ASEAN) chartered professional engineer. He received his BEng (Hons) in Electrical and Electronics Engineering from Universiti Teknologi Petronas in 2004, MSc in Microelectronics from Universiti Kebangsaan Malaysia in 2005, and Ph.D. from Universiti Tunku Abdul Rahman in 2011. In 2008 and 2015, respectively, he underwent research attachments at the University of Oxford, UK, and Nippon Institute of Technology, Japan. He has published more than 180 research articles, including patents, refereed journal papers, conference proceedings, books, and book chapters.

Contents

Preface

Cosmology is the scientific study of the universe, that is, its origins, structure, and ultimate fate. Most specialists consider cosmology a branch of astrophysics, particularly since modern cosmology's emergence following Einstein's introduction of the general theory of relativity in 1915. This revolutionary period sparked profound questions about the workings of the cosmos and the fundamental nature of the universe.

At the turn of the 20th century, the prevailing model, known as Kapteyn's universe, depicted the cosmos as a single "island universe" containing all known celestial objects, including stars, planets, and nebulae (which were believed to be clouds of gas, like hydrogen and helium, and dust). However, groundbreaking observations by Edwin Hubble and his collaborators soon revealed the dynamic nature of the cosmos. They discovered that distant galaxies were moving away from one another, a phenomenon that became known as the *expansion of the universe*. This observation marked a pivotal moment in cosmology, laying the foundation for what is now known as the Big Bang theory.

The current cosmological model, known as Lambda-Cold Dark Matter (Λ-CDM), describes an expanding universe that began in a hot, dense state and has since evolved into the cooler, more diffuse cosmos we observe today. This model also predicts several important phenomena, including the Cosmic Microwave Background (CMB) radiation, the abundance of the lightest chemical elements, and the formation of the universe's large-scale structure, such as galaxy clusters and superclusters. It further explains how stars, planets, and potentially life formed within galaxies.

This volume is a collection of essays that explore various aspects of cosmology, providing a comprehensive understanding of the universe and its workings. The editors hope that readers will be inspired to explore cosmology and further investigate the mysteries of the cosmos.

This volume begins with an introduction by Kim Ho Yeap and Hui Chieh Teoh, which offers a panoramic review of contemporary cosmology. Following this is Chapter 2, "Life in the Cosmos: Paradox of Silence and Self-Awareness" by Jonathan H. Jiang, Avery M. Minion, and Stuart F. Taylor. Chapter 3, "What Is Spacetime? How Was It Born?" by Sho Fujita, is a pedagogical chapter exploring the nature of spacetime and its origin. John C. Botke contributes Chapter 4, "Cosmology with Time-Varying Curvature: A Summary" while Herwig Dejonghe explores rotating cosmological models in Chapter 5, "Rotating Cosmological Models, or How Matter and Light Can Escape from a Rotating Black Hole". Michael G. Albrow presents Chapter 6, "Closed Loop Universe and the Topology of Totality". The volume concludes with Chapter 7,

"Non-Particulate Quantum States of the Electromagnetic Field in Expanding Space-Time" by Philip Broadbridge, Sarah Becirevic, and David Hoxley.

Sincerely yours,

Herman J. Mosquera Cuesta (Ph.D. Relativistic Astrophysics)
Colciencias Visiting Scientist,
National Program of Basic Sciences,
Space Science Division,
Bogotá, Colombia

Valencia International University (VIU),
Astronomy and Astrophysics Master Science Program,
Valencia, Spain

Acknowledgements

The Academic Editor of this volume (HJMC) would like to sincerely recognise the invaluable support received from the IntechOpen Publishing Process Managers Ms. (Mrs.) Viktorija Zgela, Ms. (Mrs.) Elvira Baumgartner and Ms. (Mrs.) Tea Jelaca. HJMC is also grateful to the IntechOpen Scientific Board and Directors for the invitation to act as the Academic Editor of this volume.

Finally, HJMC would like to also thank the Mr. Pacho LAN internet site Manager César Augusto Chavarría A. and to the LAN operator Durany Laverde for their hospitality and technical assistance during the preparation of this book while being in Medellín, Colombia, South America.

Section 1

Introduction

Chapter 1

Unraveling the Mystery of the Cosmos: A Review

Kim Ho Yeap and Hui Chieh Teoh

Abstract

Cosmology is a discipline that describes the physical nature of the universe. This discipline encompasses the branch of physics, metaphysics, and astronomy. In general, the universe is believed to start from the Big Bang, followed almost instantaneously by the cosmic inflation approximately 13.8 billion years ago, and it is conjectured to end, very likely, with the Big Freeze. Cosmologists are interested on the origin, evolution, and the ultimate fate of the universe. This chapter brings readers to take a glimpse on the mystery of the universe. It gives a succinct elucidation on the growing journey of the universe—from its birth, its advancement to the present and future, to its demise. The technology used to detect cosmic sources, the theories that underpin the universe, and some of the equations used to estimate its properties are also briefly illustrated here.

Keywords: big bang, big crunch, big freeze, big rip, cosmic microwave background (CMB), cosmological constant, Hubble's law, inflation, lambda-CDM model, radio telescope, theory of everything, universe

1. Introduction

Imagine reposing yourself on a couch in your backyard after pandering your appetite with a palatable dinner. You raised your head and looked up onto the sky. It was a clear fine night, with the round white moon and twinkling stars illuminating the sky. A sudden surge of adrenaline rushed through your bloodstream, when the epiphany of how nature itself could create such a breath-taking scenery struck you. You immersed yourself into the beauty of nature. Then, all of a sudden, you realized that something looked different—one of the stars inclined toward the north was behaving in a peculiar fashion. It was glowing in bluish-white and was much brighter than any other stars, and, above all else, you could not recall it being there in the previous nights. You were baffled and started to sink yourself into deep contemplation. You were searching for an answer for its sudden emergence, but as a novice who knows next to nothing about cosmic elements in outer space, you failed to find a convincing one. Coincidentally, an astronomer, an astrophysicist and a cosmologist were staring at the sky, and they too noticed that bluish-white star. As aficionados in their fields, however, they could probably deduce at first sight that it was the remnants of a dying massive star, known as a supernova. Had you been equipped with the knowledge beyond the atmosphere of earth, you too would be able to draw the same conclusion as well.

Now, we know that astronomers, astrophysicists, and cosmologists study astronomy, astrophysics, and cosmology, respectively, and that these three disciplines deal with celestial elements and phenomena, but how different are they from each other? Whilst there are overlapping areas among these three fields and that they symbiotically assist each other, their differences are indeed rather distinctive and should not be confused. In general, both the astrophysics and cosmology disciplines are branches sheltered under the large astronomy umbrella.

Astronomy can be broadly defined as a discipline that studies celestial phenomena and the physical and chemical properties of cosmic bodies [1], such as planets, asteroids, comets, stars, nebulae, supernovae, galaxies, quasars, and black holes, just to name a few. It is considered as one of the oldest fields of natural science, in which its history could be dated back to as early as 5000 years ago, when the ancient Egyptians studied the stars with their naked eyes. One of the oldest artifacts of astronomical observation is the Nebra sky disk, which is believed to have been constructed in the 1600s BC [2]. As can be seen in **Figure 1**, this 30-cm disk, found in northern Europe, is engraved with images of the sun, crescent moon, and stars. Although the initial intention of this piece of art is still a mystery, there are a few interpretations, with some people believing it to be a sky map. Today, the positions of celestial objects, including those that are invisible to our naked eyes, can be mapped by astronomers using sophisticated instruments and telescopes, and they endeavor to accurately plot the positions and movements of these objects. Observations through modern telescopes and modeling through mathematical techniques also enable astronomers to better understand the celestial objects and their interactions with the surrounding environment. These techniques are the two main types of modern astronomy, that is, observational astronomy and theoretical astronomy. As their names imply, the distinction between these two types of astronomy is obvious—observational astronomy identifies the celestial bodies and phenomena via experimental inspection; whereas theoretical astronomy uses mathematics and physics to describe the observed objects and phenomena to tell their past, present, and future.

Figure 1.
The Nebra sky disk. Credit: Vincentz, 2022 (https://commons.wikimedia.org/wiki/File:Nebra_1b.jpg).

Strictly speaking, astrophysics could be perceived as a conglomeration of both observational and theoretical astronomy. This discipline is relatively new—it was only coined as such in the nineteenth century when the German physicist, Joseph Fraunhofer invented and used the spectroscope to study the spectra of cosmic sources. Astrophysicists attempt to learn about the characteristics (such as their chemical properties and energy density) of cosmic objects and phenomena and to study how these objects behave and have evolved over time by means of physics, chemistry, and mathematics. They apply the theories of relativity, quantum mechanics, particle and nuclear physics, thermodynamics, and other fields to better understand what they observe. For example, they might study how a protostar developed into a massive star and finally ended its life as a black hole or pulsar. Given its intricate complexities, nowadays, most works performed in astronomy involve astrophysics; thus, both the terms are used interchangeably today.

Cosmology, on the other hand, differs from astrophysics in that it studies the universe in its entirety rather than individual cosmic sources. Cosmologists endeavor to find answers for the universe's origin, its expansion to its present state, and its ultimate fate. They also attempt to understand the laws that keep the universe in order and study its properties on a larger scale, including cosmic microwave background radiation (CMBR), dark energy, and dark matter. Cosmologists employ various theories in their work, including the theories of relativity, quantum field theory (QFT), electromagnetic theory, quantum chromodynamics (QCD) theory, quantum electrodynamics (QED) theory, string theory, and the M-theory. Some of the current topics that cosmologists are exploring include the multiverse concept, the theory of everything, the validity of the M-theory, and the possible fates of the universe (the Big Crunch, Big Rip, and Big Freeze). Cosmologists often work closely with astronomers and astrophysicists, as they are required to analyze the data obtained from cosmic observations.

This book focuses on the past, present, and future of the universe and presents some of the latest discoveries that have been puzzling both the public, in general, and scientists, in particular. The first chapter gives a succinct elucidation on the growing journey of the universe—from its violent birth to its current state and future demise. It also provides an overview of the forces that govern nature and the theories that are used to describe them. Additionally, the chapter discusses the technology used to detect cosmic sources that enable us to better understand the universe, including the cosmic microwave background radiation (CMBR) and the relevant theories used to calculate the universe's properties, such as its size during different phase transitions.

2. The birth of the cosmos: The big bang model

In cosmology, there are two popular postulates to explain the origin of the universe. They are the steady-state and the Big Bang models. In the steady-state model, cosmologists believe that matter is constantly created as the universe expands, resulting in a constant density of matter over time [3]; the Big Bang model, on the contrary, advocates the expansion of the universe from the singularity, where the space curvature, density, and temperature were infinite at a point in time, and this results in a progressive dilution of matter [4]. There was initially a comparable number of supporters for both models. The Big Bang model, however, has gradually gained more traction, overtaking the steady-state model, as further observational pursuits evince an increasingly number of evidence toward it.

One of the observations, which convinced astrophysicists about the validity of the Big Bang model, is the red shifts of galaxies based on the Doppler effect. According to the Doppler effect, the frequency of a moving object T is perceived to be higher when it is approaching a stationary observer R at velocity + u, and conversely, the observed frequency is lower when T is moving away from R at velocity -u. The Doppler effect can be approximated as [5],

$$f_R \approx f_T \left[1 + \frac{(\pm u)}{c} \cos\theta\right] \tag{1}$$

where c $\simeq 3.0 \times 10^8$ m/s is the speed of electromagnetic signal at free space, θ the angle of T relative to the direct line to R (i.e., $\theta = 0$ when both T and R lie on the same plane), and f_R and f_T denote, respectively, the frequency of the signal perceived by R and that transmitted by T. The fact that the observed frequencies of galaxies shifting toward the lower frequency end of the electromagnetic spectrum indicates that they are receding from Earth. By assessing how the spectra of galaxies were being red-shifted in the 1920s, American astronomer, Edwin Hubble found that all galaxies were moving away from Earth and that the farther they were from Earth, the higher were their recession rates [6, 7]. This implies that the universe is indeed expanding, as predicted by Einstein's original general theory of relativity. The observation leads to Hubble's law, which concludes that the velocity of a distant galaxy v_R is directly proportional to its distance from earth d_g and is expressed as,

$$v_R = H_0 d_g \tag{2}$$

where H_0 is the Hubble parameter at the present time and is written in the form of $H_0 = 100\,h$ kms^{-1}Mpc^{-1}, with the constant h parameterizing the uncertainty. Planck measurements of the cosmic microwave background (CMB) gives h $\simeq 0.674 \pm 0.005$ [8, 9].

Another evidence that further corroborates the Big Bang model is the cosmic microwave background radiation (CMBR). Discovered serendipitously in 1965 by American astronomers, Arno Penzias and Robert Wilson [10], the CMBR is the remnant that survived from the early stage of the universe when it was only 375,000 years old [11]. It conceals a plethora of spatial and spectral information about the infant universe. At 375,000 years after the Big Bang, the CMBR possessed a black body temperature of 3000 K. Owing to the cosmic expansion, however, the temperature has dropped to 2.725 K at the present, and it peaks at the microwave region of the electromagnetic spectrum.

By observing the CMBR, astrophysicists found various puzzling phenomena. Firstly, they learned that the temperature of the CMBR is virtually uniform across the entire sky. Now, other than time dilation and length contraction, we learn from Einstein's special theory of relativity that no information can travel faster than the speed of light c, as given in (3)–(5),

$$E = \frac{mc^2}{\sqrt{1 - \left(\frac{v}{c}\right)^2}} \tag{3}$$

$$p = \frac{mv}{\sqrt{1 - \left(\frac{v}{c}\right)^2}} \tag{4}$$

$$E^2 = (pc)^2 + (mc^2)^2 \tag{5}$$

where E is the energy of an object with mass m, momentum p, and velocity v. This is to say that two causally disconnected regions are unlikely to share the same temperature. Astrophysicists refer to this problem as the horizon problem.

The second puzzling phenomenon is the flatness problem. Measurements of the anisotropies in the CMBR show that the curvature constant k is nearly zero, which implies that the universe is practically flat. To astrophysicists, this finding was somewhat out of their expectation. Since Einstein's general theory of relativity states that mass warps the fabric of space-time, the universe, which consists of mass, is naturally predicted to have a certain curvature. The curvature constant k is given in (6) as [12].

$$k = \left(\frac{HR}{c}\right)^2 \left(\frac{\varepsilon}{\varepsilon_c} - 1\right) \tag{6}$$

where H is the Hubble parameter, R the radius of the universe, and ε the energy density. The critical density ε_c represents the energy density for which the geometry of the universe is flat, and it is given by [9].

$$\varepsilon_c = \frac{3H^2}{8\pi G}. \tag{7}$$

where G is Newton's gravitational constant.

To form celestial objects such as stars and galaxies through gravitational effect, anisotropies must exist in the early universe. When CMBR was initially discovered, it appeared to be uniformly distributed in all directions. The launch of the Cosmic Background Explorer (COBE) satellite in 1989 probed more extensively into the CMBR, and it unveiled small random fluctuations of temperature throughout the sky [1]. In 2009, the European Space Agency (ESA) launched the Planck space telescope, which was capable of distinguishing temperature variations in the CMBR in the order of one part in a million [13]. **Figure 2** illustrates the tiny fluctuations of the CMBR as

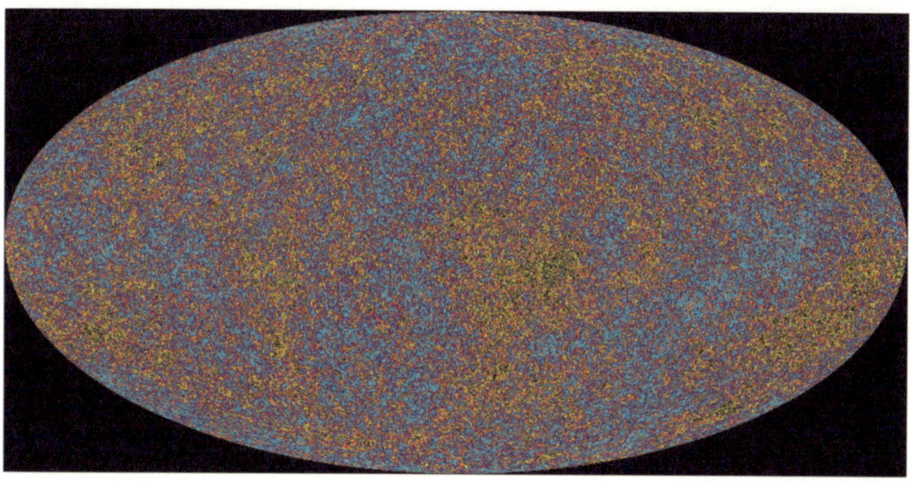

Figure 2.
Anisotropies in the CMB. Credit: Imagtek, 2019 (https://commons.wikimedia.org/wiki/File:PlanckMCD256H.png).

seen by Planck. Although the observational results confirmed the prediction of anisotropies in CMBR, astrophysicists were uncertain as to how they could exist, and this becomes the third problem of CMBR.

In 1981, American cosmologist Alan Guth brought forward the idea of cosmic inflation, which can help to solve the problems of CMBR [14]. The theory was later fine-tuned independently by Andrei Linde and by Andreas Albrecht and Paul Steinhardt. According to the cosmic inflation theory, when the nascent universe cooled down, it was trapped in a false vacuum, which constituted high energy density. This positive energy mediated an exponential expansion of space based on Einstein's general theory of relativity. The accelerating expansion, which was faster than the speed of light, was believed to have started at 10^{-36} s and ended at 10^{-33} s to 10^{-32} s after the Big Bang. Astrophysicists named the field and quanta, which drove the cosmic inflation, the inflaton. Friedmann modified Einstein's equation of general relativity and was able to show that the radius of the universe R can be expressed as a function of time t, that is, [12],

$$R = R_0\, exp\left(\sqrt{\frac{2\Lambda}{3}}t\right) \tag{8}$$

where R_0 is the radius of the universe at t = 0, and Λ is the cosmological constant, which is computed to be $2.036 \times 10^{-35}\ s^{-2}$ [15]. It can be easily seen that the size of the universe expanded 10^{26} times in 10^{-34} s [12].

The inflation solves the horizon problem because all points were causally in contact prior to it. The flatness problem can also be explained by the theory since the rapid expansion stretches the space curvature of the universe to almost flatness. During inflation, the quantum fluctuations in the density of matter were enlarged to astronomical scale. Hence, the anisotropies in the CMBR are also deemed sensical. Besides solving the three problems observed from the CMBR, the inflation theory also answers the monopole problem. This problem arose because physicists failed to find traces of topological defects such as magnetic monopoles. Based on the grand unified theories (GUTs), however, there should be a myriad of massive particles and topological defects when the symmetry of grand unification was broken. After the inflationary epoch, the density of these monopoles was diluted incredibly, rendering them extremely difficult to be detected.

In 1998, the Supernova Cosmology Project team led by American astrophysicist Saul Permutter [16] and the High-Z Supernova Search team led by American astrophysicist Adam Riess and Australian astrophysicist Brian Schmidt [17] measured the universe expansion using Type Ia supernovas as standard candles. To their astonishment, they found that the high-redshift supernovas were fainter than expected, an indication that the expansion rate has been accelerating. Astrophysicists attribute the acceleration to the repulsive force imposed by dark energy. Although dark energy is still an unknown form of energy to us, we have various explanations for it. For instance, some physicists hypothesized that it is a kind of dynamic field known as quintessence. The simplest possible explanation for dark energy, however, is that it is a quantum vacuum energy of the sort, which could be mathematically associated with the cosmological constant Λ in Einstein's general theory of relativity in (9)

$$G_{\mu\upsilon} + \Lambda g_{\mu\upsilon} = \frac{8\pi G}{c^4}\left(T_{\mu\upsilon}{}^{mat} + T_{\mu\upsilon}{}^{vac}\right) \tag{9}$$

where $G_{\mu\upsilon} = R_{\mu\upsilon} - \frac{1}{2}Rg_{\mu\upsilon}$ is the Einstein's tensor, which depicts the space curvature; $R_{\mu\upsilon}$ the Ricci curvature tensor; R the scale curvature; $g_{\mu\upsilon}$ the metric tensor, which constitutes the properties of every point within the space-time fabric; $\frac{8\pi G}{c^4}$ and G denote, respectively, Einstein's and Newton's gravitational constant; and $T_{\mu\upsilon}{}^{mat}$ and $T_{\mu\upsilon}{}^{vac} = \rho_{vac}g_{\mu\upsilon}$ are, respectively, the energy-momentum tensor, which represents the local energy, momentum, and stress within the space-time fabric for ordinary matter and vacuum. The left-hand side (LHS) of (9) describes how matter-energy warps space-time, whereas the right-hand side (RHS) of the equation describes the propagation of matter-energy through space-time. Upon correlating the LHS and RHS of (9), one can find that vacuum energy and Λ are identical when the vacuum energy density ρ_{vac} (is)

$$\rho_{vac} = \frac{\Lambda c^4}{8\pi G}. \quad (10)$$

It is worthwhile noting that one primary outstanding issue, which stems from this relationship, is that the theoretical value of Λ computed from QFT mismatches the value obtained from the study of the Type Ia supernovas, with the former deviating from the latter by more than the order of 120 [18]. This issue is commonly known as the cosmological constant problem, and it is still up in the air, awaiting an answer today.

3. The evolution of the cosmos

The current standard model used to explain the universe is known as the Lambda Cold Dark Matter or, in short, Lambda-CDM or ΛCDM model. As implied by its name, the universe is regarded as comprising three major components – viz., the dark energy (denoted by the cosmological constant Λ), the cold dark matter, and the ordinary baryonic matter. The cold dark matter is an invisible matter that interacts weakly with the ordinary baryonic matter. Astronomical observations and cosmological theory suggest that the universe is made up of 69% of dark energy, 25% of dark matter, 5% of "ordinary" atomic matter, and 1% of other observable components. Among the remaining 1% components, neutrinos, CMBR, and black holes constitute, respectively, 0.1%, 0.01% and 0.005% [19]. The breakdown of the composition of the universe is graphically summarized in **Figure 3**.

According to the ΛCDM model, the Big Bang commenced when the universe expanded from the singularity. Based on the timeline after the Big Bang, the evolution of the universe can be broadly classified into the Planck (0 to 10^{-43} s), grand unification (10^{-43} to 10^{-36} s), inflationary (10^{-36} to 10^{-32} s), quark (10^{-12} to 10^{-5} s), hadron (10^{-5} to 1 s), lepton (1 to 10 s), nucleosynthesis (10 to 1000 s), recombination (375,000 years), dark age (375,000 to 200 million years), star and galaxy formation (200 million to 1 billion years), solar system formation (9.2 billion to 9.3 billion years), Stelliferous (10^{6} to 10^{14} years), degenerate (10^{15} to 10^{39} years), black hole (10^{40} to 10^{100} years, assuming that proton decay occurs), and dark (10^{100} years onwards, assuming that proton decay occurs) eras [20]. A summary of the evolution of the universe from the Big Bang until the present is given in **Figure 4**.

In the Planck era, the size of the universe is merely about 10^{-35} m, and it is dominated by the quantum effect of gravity. When it entered the grand unification

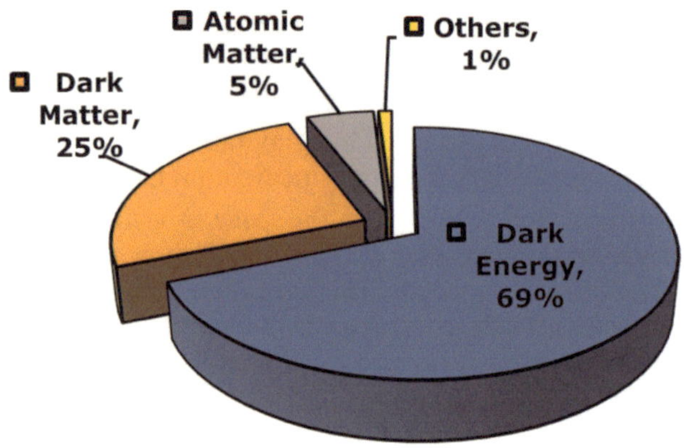

Figure 3.
The components that compose the universe.

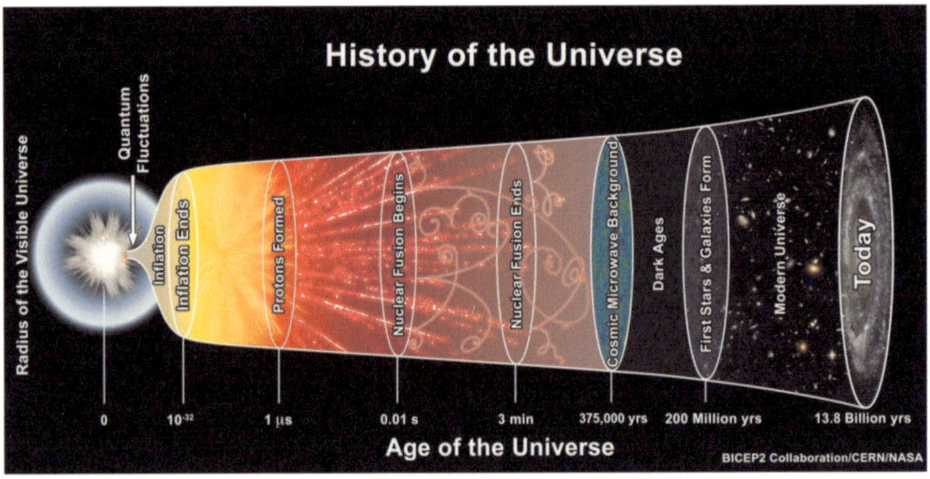

Figure 4.
Timeline of the universe. Credit: TheAstronomyBum, 2014. (https://commons.wikimedia.org/wiki/File:The_History_of_the_Universe.jpg).

era, gravitational force started to separate from the other three forces. During the inflationary epoch, the size of the universe underwent an exceedingly rapid expansion in a very brief period of time. Since then, both the temperature and density of the universe have been dropping. The strong nuclear force separated out from the electroweak force during the cosmic inflation. The weak nuclear force subsequently separated from the electromagnetic force in the quark era, resulting in the presence of the four fundamental forces of nature today. Baryons such as protons and neutrons and their anti-particles were formed from quarks and anti-quarks in the hadron era. Leptons, including electrons and their corresponding neutrinos started to dominate in the lepton era. During the nucleosynthesis period, protons and neutrons combined to form nuclei of simple atoms, that is, hydrogen, helium, and lithium, via nuclear fusion. In the recombination era, electrons were attracted into the orbits of hydrogen and helium nuclei, forming neutral atoms. It was also in this era that the photons from the cosmic microwave background became disentangled from the charged particles,

DOI: http://dx.doi.org/10.5772/intechopen.1001923

thereby, ensuing a transparent universe. The period after the recombination era and before the formation of quasars, stars, and galaxies is better known as the dark age because the universe is literally dark in this window. About 200 million years after the Big Bang, the gravitational attraction due to the irregularities in the universe cause molecular gas of hydrogen to coalesce to form protostars and then stars. This was followed by the attraction of stars into clusters, mediating the formation of galaxies about 1 billion years after the Big Bang. Our solar system was formed relatively late, roughly about 9.2 to 9.3 billion years after the Big Bang. Today, we are living in the Stelliferous era, where the universe has existed for about 13.8 billion years. This era is predicted to last for a maximum of 10^{14} years after the Big Bang. Subsequently after that, in the degenerate era, all stars have used up their hydrogen fuel. In this period, the universe is dominated by degenerate remnants made up of brown dwarfs, white dwarfs, neutron stars, and black holes. Brown dwarfs, white dwarfs, and neutron stars eventually will be gone too, as a result of proton decay. What will be left in the universe are only black holes. The universe therefore enters the black hole era. But when all the black holes evaporate via Hawking radiation, the universe becomes virtually empty. This is the dark era where the universe is merely dominated by electrons, positrons, and dark matter at this point. In this one possible fate, the universe will face its Big Freeze where it finally rests peacefully at an extremely low energy level, with its temperature settled slightly above absolute zero.

4. The fate of the cosmos

The universe will ultimately reach an end at some point. Nevertheless, there is no one common agreement as to how it will end. Before the discovery of the dark energy in 1998, many believed that there may be a point in time where the density of the universe exceeds the critical density. If this occurred, the universe would cease from expanding, and instead, it would start collapsing back into the infinitely dense singularity. This phenomenon is coined the Big Crunch. A new Big Bang may occur again after the Big Crunch. Nevertheless, most cosmologists decided to ditch this idea now, realizing that dark energy is accelerating the expansion rate of the universe. Hence, it is more likely that the universe will encounter either the Big Rip or the Big Freeze. In the Big Rip scenario, dark energy will eventually outvie gravitational force, and it becomes incredibly strong, so much so that the universe is torn apart. A more plausible and popular prediction, however, is the Big Freeze theory (also known as the Big Chill and the heat death). According to this theory, the universe will ultimately reach thermodynamic equilibrium (i.e., maximum entropy), where heat is fairly distributed. Since energy distribution is homogeneous, no further work is possible. The universe eventually comes to its quiet and lonely death [21].

5. The forces of nature

Four fundamental forces are known to exist in nature. They are the strong nuclear force, weak nuclear force, electromagnetic force, and the gravitational force [12, 22].

i. The strong nuclear force is the strongest among the four forces. It is known to bind hadron particles – viz. protons and neutrons, within the nucleus; it also confines quarks together to form protons and neutrons. The strong nuclear force is the energy source released from nuclear fission and fusion. The strong interaction has a range of 10^{-15} m. According to the quantum chromodynamics (QCD) theory, the strong force is caused by the exchange of the gluon.

ii. The weak nuclear force is in the order of 10^{-14} that of the strong force. The weak force acts in each individual nucleon (which is the collection of protons and neutrons). This force is responsible for the radioactive beta decay of the atomic nucleus, that is, in beta-minus decay, a neutron turns into a proton, an electron, and an electron antineutrino; whereas in beta-plus decay, a proton is turned into a neutron, a positron, and an electron neutrino. The weak interaction has a range of 10^{-17} m. The weak force is given by the exchange of the W^+, W^-, and Z gauge bosons.

iii. The electromagnetic force is in the order of 10^{-2} that of the strong force, and it is responsible in binding atoms together. This force also governs the interaction among electrically charged particles—when particles with identical charges (i.e., both of them are either positively or negatively charged) approach each other, they tend to repel, while those with opposite charges attract each other. The electromagnetic field consists of both electric and magnetic fields. During static condition, the fields exist independently. When the fields vary with time, however, they have to exist simultaneously [22]. The electric force $\mathbf{F_e}$ exerted on a charge $q = 1.6 \times 10^{-19}$ C is given by

$$\boldsymbol{F_e} = q\boldsymbol{E} \tag{11}$$

where **E** is the intensity of the electric field experienced by q. Likewise, the magnetic force $\mathbf{F_m}$ is given by

$$\boldsymbol{F_m} = q\boldsymbol{u} \times \mu \boldsymbol{H} \tag{12}$$

where $\times$ is the cross product, **u** the velocity vector, μ the permeability of the material, and **H** is the magnetic field intensity. The range of the electromagnetic interaction is infinity. The magnetic field is especially important when studying the characteristics of cosmic sources. According to the quantum electrodynamics (QED) theory, photons carry the electromagnetic force.

iv. The gravitational force is the weakest among all forces. It is in the order of 10^{-39} of the electromagnetic force. The gravitational force attracts any objects with mass. Hence, when the object comprises a large mass (for e.g., a star), gravitational force outweighs other forces. Based on Newton's law, gravitational force F_g can be expressed as

$$F_e = \frac{Gm_1m_2}{r^2} \tag{13}$$

where $G = 6.674 \times 10^{-11}\ m^3kg^{-1}\ s^{-2}$ denotes Newton's gravitational constant and r the distance between two objects with masses m_1 and m_2. The range of gravitational interaction is infinity. The gravitational force is caused by the exchange of a quantum, known as the graviton.

DOI: http://dx.doi.org/10.5772/intechopen.1001923

6. The theory of everything

Physicists have attempted to explain the forces using quantum field theory (QFT). The first force successfully described by QFT was the electromagnetic force. Known as the quantum electrodynamics or QED, this theory suggests that photons act as the carriers of the electromagnetic force and that the force is transmitted by the exchange of these gauge bosons. This is to say that the electromagnetic energy is transferred when a photon is emitted from an electrically charged particle and absorbed by the other. As can be seen from the Feynman diagram in **Figure 5**, when an electron (e^-) hits another, a photon (γ) is exchanged. The electron, which emits the photon, recoils in space, while the electron, which absorbs the photon's energy and momentum, deflects correspondingly.

Physicists found that the electromagnetic force can be consolidated with the weak force. This unified force is named the electroweak force. The theory of the electroweak force introduces the W^+, W^-, and Z gauge bosons as the quanta of the weak field. The strong force, on the other hand, can be explained using quantum chromodynamics (QCD). According to the QCD theory, baryons (i.e., proton, neutron, delta, lambda, sigma, xi, and omega) and mesons (i.e., pion, eta, kaon, phi, psi, rho, upsilon, d, and b) are composed of quarks. The exchange of the quantum gluon mediates the strong interaction between two quarks.

The successful establishment of a single theory, which combines both the weak and electromagnetic forces, has motivated physicists to explore feverishly for a single theoretical framework of physics capable of explaining all physical phenomena in the universe. Being coined the theory of everything (TOE), this single theory should be able to effectively relate the strong interaction, weak interaction, electromagnetism, and gravitation. The dream of realizing the TOE, however, has not been smooth sailing. Physicists soon learned that such dream is but tantalizing.

In their subsequent step toward the TOE, physicists endeavored to describe the electroweak force and the strong force as a single force. Albeit their persistent effort in materializing this grand unified theory (GUT), luck has not been on their side. Merging the electroweak theory and the QCD theory has not been successful hitherto. Alas, physicists decided to resort to an ad hoc model, which constitutes both the electroweak and QCD theories. More colloquially referred to as the standard model today, this model has been found to produce results, which agree closely with the observational evidence. Even so, however, physicists are not quite contented with the fact that they have to treat the electroweak and strong forces separately and not as a single unified force.

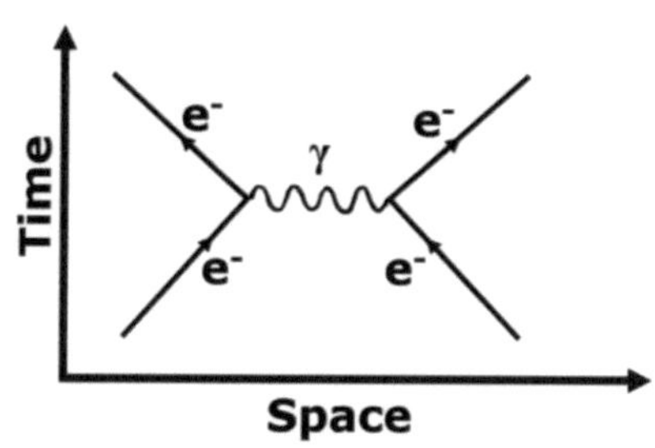

Figure 5.
A Feynman diagram showing two electrons exchanging a photon, when one collides with the other.

Perhaps the problem that frustrates physicists the most is not their failure in establishing the GUT but their futile efforts in accounting for the gravitational force. As compared to the latter, the former is nothing more than a mere minor setback. The gravitational force which is depicted by Einstein's general theory of relativity—a theory that describes how gravity mediates the curvature of space and time on a larger scale— that is essentially different from the quantum theory. Owing to the Heisenberg uncertainty principle and the probabilistic nature of the quantum theory, physicists find it particularly difficult to quantize gravity. Without the quantum theory of gravity, gravitational force could not be fitted into the standard model, let alone the theory of everything. Above all else, the infinitesimal world of particles could not be reconciled with the infinitely large scale of the universe.

It was not until the development of the supergravity theories that the hope for the realization of the TOE was rekindled. Being a type of quantum field theory, these theories relate general relativity with supersymmetry at the subatomic level. Supergravity predicts gravitational force to be carried by a gauge boson, known as the graviton, and based on the concept of supersymmetry, it is hypothesized to have a superpartner named the gravitino. Supergravity is regarded as the low-energy limit of the higher dimensional string theory. The string theory offers a paradigm shift in the structure of the fundamental constituents of nature. As its name indicates, string theory depicts particles in the form of tiny pieces of vibrating strings, instead of points. Another bizarre feature of the string theory is that it predicts the existence of 10 dimensions, that is, nine spatial and one temporal dimensions. Many believe that six of the dimensions are being curled up or "compactified" into the internal spaces, which are of sub microscopic scales, such that only four dimensions are left noticeable (viz the height, width, length, and time dimensions that one is familiar with). The extra dimensions pose the advantage of allowing supergravity to meld gravity with the other three forces—namely, the electromagnetic and the strong and weak nuclear forces.

A predicament that physicists faced was that they found five consistent, but seemingly different, string theories could be derived to explain how the extra spaces could be compactified. The five theories are the Type I, Type IIA, Type IIB, Heterotic SO32, and Heterotic E8 theories. Physicists had difficulty identifying which version of the theory was the genuine one and which four were phony. Further investigations on the theories, nevertheless, suggested that these five theories could be the same after all—that they end up describing the same phenomena in different manners. Today, physicists believe that the string theories (including supergravity) are mere subsets of the more rudimentary M-theory. This theory predicts that 11, instead of 10, dimensions exist and that particles are made not just from strings but objects of different dimensions. The term p-branes is used to define these objects, with the letter p representing integers ranging from zero to nine. It is worthwhile noting that M-theory inspires the multiverse concept—an idea that suggests that different parallel universes are present, with each one of them governed by different laws of physics. Many, however, still cast doubt on the validity of the M-theory, and not to mention the string theory and supergravity as well, since evidence are yet to be found to support them.

7. Cosmic detection

As mentioned in the preceding section, data collection on cosmic sources is crucial for the study of cosmology. The data are obtained through the use of telescopes, which are

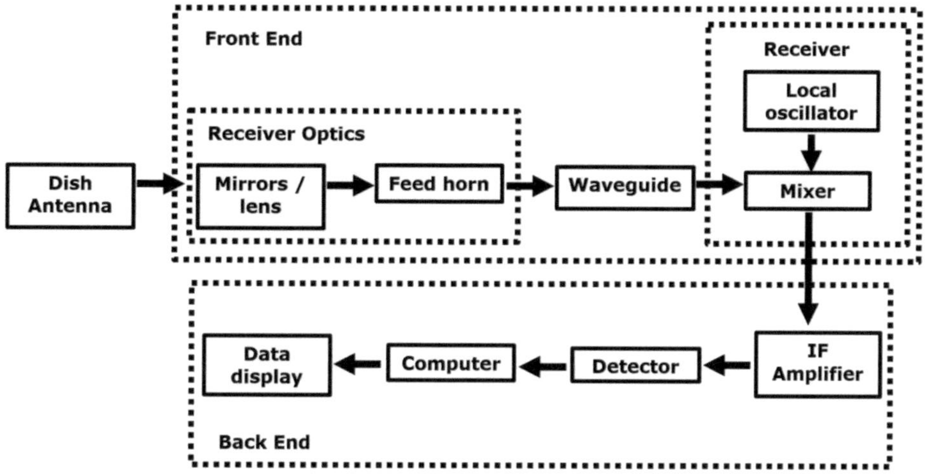

Figure 6.
Block diagram of a radio telescope.

specifically designed to detect and collect electromagnetic signals emanating from cosmic sources. Due to the cosmic expansion, the cosmic sources in the early universe had been stretched into the radio region of the electromagnetic spectrum. Thus, radio telescopes play a critical role in the study of cosmology, providing cosmologists with the data necessary to understand the universe and its evolution. As illustrated in **Figure 6**, a typical modern radio telescope, such as those used in the Atacama Large Millimeter/submillimeter Array (ALMA) interferometer, constitutes the following components [23, 24].

i. A dish antenna, which consists of a main and a sub-reflector. Dish antennas are used in this case because they have the highest aperture efficiencies when compared with the other types of antenna configurations. The function of the antenna is to detect electromagnetic signal radiated from cosmic sources. The most common types of dish antennas used in radio astronomy are the Cassegrain and Gregorian offset configurations [25–27].

ii. A receiver optics, which is usually constructed from either a pair of mirrors (in some rare occasions, a lens is used, instead) and horns or a horn and an orthomode transducer (OMT). The purpose of the receiver optics is to focus the signal to the feed horn. An example of the receiver optics for the ALMA telescope is shown in **Figure** 7 [24, 28].

iii. A waveguide, which is used to channel the signal from the feed horn to the mixer circuit [29–31]. Waveguide can either be circular or rectangular.

iv. A receiver circuit, which is used to feed the signal to the electronics. In radio astronomy, coherent receivers are used to detect the phase and intensity of the signal. Incoherent receivers, on the other hand, only accounts for the intensity of the signal.

v. A detector circuit, that is, a non-linear square-law detector is used in a radio telescope to demodulate the down-converted modulating signal from its carrier.

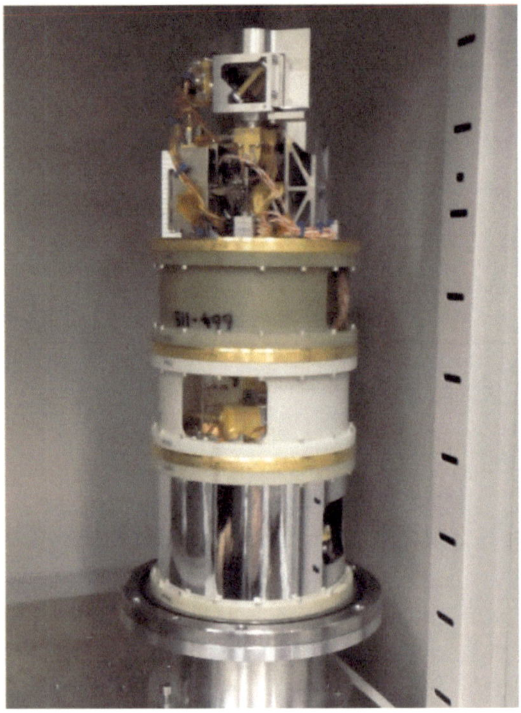

Figure 7.
The band 8 receiver optics fitted into a self-contained cartridge for the ALMA telescope.

vi. A spectrometer, which allows the spectrum of the signal to be viewed and analyzed.

To understand the operation of a radio telescope, one can imagine the beam from a cosmic source to be collected by the dish antenna. The main reflector scatters the beam to the sub-reflector. The sub-reflector, in turn, illuminates the first and then the second mirrors. The signal is subsequently focused onto the feed horn. The waveguide connected to the horn channels the signal to the receiver circuit. The mixer circuit at the receiver modulates the signal with another lower frequency signal generated from a local oscillator (LO). The process of modulation down converts the signal to a lower intermediate frequency (IF) signal. This IF signal is then fed to a detector circuit after it goes through amplification. The detector circuit rectifies the signal before sending it to a spectrometer for analysis. By analyzing the electromagnetic signal emitted by cosmic sources, cosmologists can gain insight into their properties and behavior. This allows us to study a wide range of cosmic phenomena, from distant galaxies and quasars to the CMBR.

8. Conclusion

A brief overview of cosmology is given in this chapter. In a nutshell, cosmology is a branch of astronomy thar studies the evolution process of the universe, from its violent birth to its death. The universe is believed to have expanded from an extremely hot and dense singularity about 13.8 billion years ago—a process that cosmologists name the Big Bang. During the early phase of its expansion, the universe underwent a cosmic inflation, which allowed it to expand to an incredibly large size within split second. The sudden fast expansion of the universe created the primordial seeds for large-scale structures. Gravitational attraction at certain regions with higher density collapses molecular clouds to form stars and galaxies. The universe is predicted to end with the Big Freeze, where it reaches the state of thermodynamic equilibrium.

Author details

Kim Ho Yeap[1]* and Hui Chieh Teoh[2]

1 Universiti Tunku Abdul Rahman, Kampar, Perak, Malaysia

2 Universiti Tunku Abdul Rahman, Kajang, Selangor, Malaysia

*Address all correspondence to: yeapkh@utar.edu.my

References

[1] Yeap KH, Hirasawa K, Nisar H. Introduction to radio astronomy and radio telescopes. In: Yeap KH, Hiarasawa K, editors. Analyzing the Physics of Radio Telescopes and Radio Astronomy. US: IGI Global; 2020. pp. 1-32

[2] Meller M. Die Himmelsscheibe von Nebra –ein frühbronzezeitlicher Fund von aussergewöhnlicher Bedeutung. Archäologie in Sachsen-Anhalt. 2002; Band;**1**:7-20

[3] Hoyle F. A new model for the expanding universe. Monthly Notices of the Royal Astronomical Society. 1948; **108**:372-382

[4] Lemaître G. The beginning of the world from the point of view of quantum theory. Nature. 1931; **127**(3210):706-706

[5] Cheng DK. Field and Wave Electromagnetics. US: Addison-Wesley; 1989

[6] Hawking S, Mlodinow L. The Grand Design. Germany: Bantam Press; 2010

[7] Hubble EP. A clue to the structure of the universe. Astronomical Society of the Pacific Leaflets. 1929;**23**:93-96

[8] Planck Collaboration. Planck 2018 results – VI. Cosmological parameters. Astronomy and Astrophysics. 2020; **641**:A6

[9] Green AM. Dark matter in astrophysics/cosmology. SciPost Physics Lecture Notes. 2022;**37**:1-32

[10] Penzias AA, Wilson RW. A measurement of excess antenna temperature at 4080 mc/s. The Astrophysical Journal. 1965;**142**:419-421

[11] Hinshaw G, Larson D, Komatsu E, Spergel DN, Bennett CL, Dunkley J, et al. Nine-year Wilkinson microwave anisotropy probe (WMAP) observations: Cosmological parameter results. The Astrophysical Journal Supplement Series. 2013;**208**(19):1-25

[12] Kisslinger LS. Astrophysics and the Evolution of the Universe. Singapore: World Scientific; 2014

[13] Teoh HC. The cosmic microwave background. In: Yeap KH, Hiarasawa K, editors. Analyzing the Physics of Radio Telescopes and Radio Astronomy. US: IGI Global; 2020. pp. 272-289

[14] Guth AH. Inflationary universe: A possible solution to the horizon and flatness problems. Physical Review D. 1981;**23**(2):347-356

[15] Carmeli M, Kuzmenko T. Value of the cosmological constant: Theory versus experiment. AIP Conference Proceedings. 2001;**586**:316-318

[16] Perlmutter S, Aldering G, Goldhaber G, Knop RA, Nugent P, Castro PG, et al. Measurements of Ω and Λ from 42 high-redshift supernovae. The Astrophysical Journal. 1999;**517**: 565-586

[17] Riess AG, Filippenko AV, Challis P, Clocchiatti A, Diercks A, Garnavich PM, et al. Observational evidence from supernovae for an accelerating universe and a cosmological constant. The Astronomical Journal. 1998;**116**: 1009-1038

[18] Hobson MP, Efstathiou GP, Lasenby AN. General Relativity: An Introduction for Physicists. Cambridge: Cambridge University Press; 2006

[19] Spergel DN. The dark side of cosmology: Dark matter and dark energy. Science. 2015;**247**:1100-1102

[20] Laughlin G, Adams F. The Five Ages of the Universe: Inside the Physics of Eternity. US: Free Press; 2020

[21] Mack K. The End of Everything: (Astrophysically Speaking). US: Scribner; 2020

[22] Yeap KH, Hirasawa K. Introductory chapter: Electromagnetism. In: Yeap KH, Hirasawa K, editors. Electromagnetic Fields and Waves. London, UK: InTechOpen; 2019. pp. 3-12

[23] Yeap KH, Hirasawa K. Fundamentals of a radio telescope. In: Yeap KH, Hiarasawa K, editors. Analyzing the Physics of Radio Telescopes and Radio Astronomy. US: IGI Global; 2020. pp. 33-55

[24] Yeap KH, Tham CY. Optimization of an offset receiver optics for radio telescopes. Journal of Infrared, Millimeter, and Terahertz Waves. 2018; **39**:64-76

[25] Yeap KH, Loh MC, Tham CY, Yiam CY, Yeong KC, Lai KC. Analysis of reflector antennas in radio telescopes. Advanced Electromagnetics. 2016;**5**: 32-38

[26] Yeap KH, Yiam CY, Lai KC, Loh MC, Lim SK, Rizman ZI. Analysis of offset antennas in radio telescopes. International Journal on Advanced Science, Engineering, Information and Technology. 2016;**6**:997-1004

[27] Yeap KH, Lai KC, Ting KC, Teh PC, Nisar H, Yeo WL. Optimization of reflector antennas in radio telescopes. Malaysian Journal of Fundamental and Applied Sciences. 2017;**13**:237-239

[28] Yeap KH, Tham CY. Receiver Optics Design for ALMA. Wireless Personal Communications. 2024;**134**:1531-1549

[29] Yeap KH, Tham CY, Yassin G, Yeong KC. Attenuation in rectangular waveguides with finite conductivity walls. Radioengineering Journal. 2011; **20**:472-478

[30] Yeap KH, Wong EVS, Nisar H, Lai KC, Ng CA. Attenuation in circular and rectangular waveguides. Electromagnetics. 2017;**37**:171-184

[31] Wengler MJ. Submillimetre-wave detection with superconducting tunnel diodes. Proceedings of the IEEE. 1992; **80**:1810-1826

Section 2

The Endless Search of the Humankind for Cosmic Contact

Chapter 2

Life in the Cosmos: Paradox of Silence and Self-Awareness

Jonathan H. Jiang, Avery M. Minion and Stuart F. Taylor

Abstract

As humanity embarks on an age of exploration, the question of whether we are alone in the universe remains unanswered. This comprehensive review reflects on the paradoxical nature of our existence in a seemingly lifeless cosmos, delving into the silence we encounter and the depths of our self-awareness. We embark on a journey that encompasses the search for life within our solar system, the mysteries of exoplanets, and the absence of technologically detectable life. Traditional definitions of life are challenged, especially in the context of artificial intelligence, as we strive to understand the complexities of existence. Contemplating our significance and insignificance in the vast cosmos, we grapple with the profound responsibility that accompanies being the only known life forms. Through introspection and contemplation, we capture the essence of our epoch—an era defined by cosmic loneliness yet magnificent self-awareness.

Keywords: extraterrestrial life, Fermi paradox, cosmic silence, self-awareness, exoplanets, artificial intelligence, human existence, cosmic exploration, existential questions, great filter

1. Introduction

At the dawn of a remarkable age of discovery, humanity finds itself in the midst of a captivating enigma that has captivated our collective consciousness: Are we the sole inhabitants of the universe? Our relentless cosmic quest has been met with an intriguing paradox—we dwell in a seemingly solitary cosmos, where definitive evidence of extraterrestrial life eludes us. As we extend our reach into the limitless expanse of space, we are confronted with a profound silence that permeates the galaxies, leaving us in awe and contemplation of our place within this vast, seemingly lifeless tableau.

To illustrate the vastness of our cosmic solitude, we need only to gaze upon the image of our tiny Earth taken from Mars, a mere speck in the vastness of the universe (**Figure 1**). This humbling perspective reminds us of the sheer magnitude of the cosmos and our seemingly insignificant presence within it. It reinforces the question that continues to drive our exploration: Are we truly alone?

However, this silence is not discouraging, nor does it diminish our pursuit. Instead, it serves as a compelling backdrop against which we illuminate the mysteries and complexities of life, both known and yet to be discovered. Standing on the precipice

Figure 1.
A photo of the earth taken by NASA's curiosity rover on 31 January 2014 from the surface of Mars. No stars are seen in this photo because the earth is shining brighter than any star in the Martian night sky. Credit: NASA jet Propulsion Laboratory-Caltech, Malin space science systems and Texas a&M University. Image source: https://climate.nasa.gov/climate_resources/89/earth-from-mars/.

of the unknown, as we peer into the cosmic void, this silence presents us with an opportunity for introspection—an invitation to delve into the very essence of our existence, to challenge our understanding, and to contemplate the multifaceted nature of life itself.

In our exploration, we navigate the intricate boundary between subjective and objective perspectives, delving into our individual and collective self-awareness. We contemplate the unique nature of our consciousness within a universe that, thus far, appears devoid of other sentient beings. This exploration transcends the confines of science, extending into the realms of philosophy and existentialism, infusing our quest for understanding with a sense of wonder and profound awe.

The silence from the cosmos is not a testament to our insignificance; rather, it beckons us to explore, to comprehend, and to marvel at our own existence. It is a call to reflect upon our position in the universe, to challenge our assumptions, and to redefine the very concept of life itself. As we venture deeper into this cosmic silence, we stand poised to rewrite our narrative, driven by curiosity, fueled by wonder, and thirsty for knowledge. This journey, propelled by our growing self-awareness and the mysteries that surround life, is an exploration not only of the universe around us but also of the universe within us.

Within the following sections of this comprehensive review, we will unravel the mysteries that arise from this paradoxical silence. We will delve into the enigmatic nature of life, exploring its origins and manifestations. We will examine the intriguing juxtaposition of proximity and paradox, investigating the potential for life within our solar system. We will turn our gaze to the distant exoplanets, beacons of hope in the cosmic night, and ponder the possibility of life beyond our own celestial neighborhood. We will venture beyond the realm of biology, redefining our understanding of life in the age of artificial intelligence. We will confront the deafening silence in the search for technologically detectable life and explore the philosophical implications of the Great Silence. We will contemplate the echoes of silence, reflecting on the absence of extraterrestrial life and its significance for our cosmic narrative. And finally, we will confront the insignificance and wonder of our place in the cosmos, recognizing our profound responsibility as the only known life forms.

Through this exploration, we seek not only to uncover the mysteries of the universe but also to gain a deeper understanding of ourselves and our place within the grand tapestry of existence. So let us embark on this extraordinary journey, where science and philosophy intertwine, where the pursuit of knowledge meets the contemplation of our cosmic significance, and where the silence of the cosmos becomes the backdrop for our profound self-awareness.

2. Unraveling the mystery of life: an interdisciplinary expedition

Life, presented in its myriad manifestations, materializes as a profound and complex enigma that weaves together elements of the biological, physical, and more recently, artificial domains. The endeavor to comprehend it constructs an image of a system that is self-perpetuating, evolving, exquisitely adaptable, and in specific instances, capable of bearing consciousness [1]. Each entity, organism, and life-form constitute a delicate thread in the vast, elaborate tapestry of existence, which evolves and transforms across eons and vast cosmic distances.

The journey to untangle the mystery of life necessitates a voyage back to the origins of our planet, Earth. Here, under the infancy of our young Sun, a symphony of primordial physical and chemical reactions played out over the course of billions of years. Within this crucible of elements, the seeds of life found fertile ground to take root, fostering the transformation of Earth from a barren rock into a thriving, living world teeming with biodiversity. Each cell, organism, individual, and species that emerged from this primeval matrix contributes a unique verse to the magnificent opus of existence [2].

However, this grand narrative is increasingly transcending the realm of the purely biological. The advent of artificial intelligence (AI) ushers us toward a new epoch, where silicon-based constructs might bear the mantle of life in addition to their carbon-based counterparts. Advanced AI systems, with their ability for autonomous learning, problem-solving, self-improvement, and potentially, self-replication, compel us to challenge and reevaluate our entrenched definitions of life. Such technological developments necessitate a paradigm shift in our understanding, pushing us to expand, and possibly even transcend, the traditional boundaries that delineate what constitutes life. As these boundary lines grow increasingly indistinct, the mysteries become deeper and our comprehension evolves [3].

Consequently, an intriguing question emerges: Could these complex constructs of code and silicon, assuming they attain a degree of consciousness, be categorized as a nascent form of life [4]? As we approach this new era, it becomes essential to reconsider the essence of life, its purposes, and our roles in shaping its trajectory [5].

Therefore, the quest to unravel the enigma of life is more than just a journey of exploration. It is equally a process of constant redefinition and conceptual evolution. As we strive to comprehend the essence of existence, tracing the trajectory of life from the biological sphere to the artificial, we delve into the profound depths of what it genuinely means to be 'alive'. This illuminating expedition into life's inception, evolution, and potential futures unravels a tale that continues to enrich our understanding of ourselves and the cosmos that encompasses us.

3. Life in the solar system: the proximity paradox

Our cosmic odyssey in search of life beyond Earth commences not within the unimaginable expanse of distant galaxies, but instead much closer to home—within the bounds of our very own solar system. Here, a diverse menagerie of celestial bodies, each a unique world in itself, performs a mesmerizing celestial ballet around our sun. Despite their relative proximity, these planetary siblings of Earth remain tantalizing enigmas, their hidden secrets shrouded within the vast, inscrutable mystery of the cosmos [6].

One particularly intriguing set of celestial bodies are the icy moons, specifically Europa and Enceladus, orbiting Jupiter and Saturn, respectively. These satellites are postulated to harbor vast subterranean oceans beneath their frozen crusts, kept liquid by tidal heating induced by the gravitational interactions with their parent planets [7]. These alien water worlds, cloaked under layers of icy armor, potentially possess the requisite conditions for harboring life akin to our terrestrial understanding of the concept. The prospect of life beneath these extraterrestrial oceans stimulates the scientific imagination and has the potential to drastically reshape our understanding of life's geographical boundaries within our solar system [8].

Turning our attention to Mars, our celestial neighbor and the focus of countless speculations and explorations, offers another fascinating dimension to this quest. This iron-rich world, whose appearance evokes images of a desert subjected to a cosmic rusting process, has long held a profound hold on our collective human imagination. The faint yet persistent signatures of methane in the Martian atmosphere, a gas commonly associated with biological processes on Earth, provide a compelling, albeit not yet definitive, clue in our interplanetary investigation for life [9].

These celestial bodies, each with their distinct characteristics, present a stark paradox. They exist relatively near to our vibrant, blue world, teeming with life in its resplendent diversity, yet their conditions seem drastically different, potentially inhospitable. As we observe our solar system from the vantage point of Earth, comparing our life-rich planet with the barren landscapes of Mars, the icy moons, and the gas giants, we are confronted with a remarkable contrast [10]. The juxtaposition of a verdant, biologically diverse Earth against the austere, seemingly lifeless backdrop of our solar system accentuates our sense of cosmic solitude, yet simultaneously fuels our intrigue.

This so-called 'Proximity Paradox' underscores the significance of our quest to determine life's cosmic prevalence, or possible rarity. Each probe we dispatch, each signal we receive, serves as a testament to our innate human desire to explore, comprehend, and ultimately discern our place in the grand cosmic narrative. Understanding this paradox could provide pivotal insights into the realms of astrobiology and cosmic biogeography, expanding our understanding of life's potential distribution in the cosmos [11].

4. Exoplanets: a beacon of hope in the cosmic night

As we journey beyond the familiar confines of our solar system, we find ourselves thrust into the magnificent amphitheater of the cosmos— an unfathomable expanse punctuated by a multitude of stars, galaxies, and the beguiling mystery of exoplanets. These remote worlds, orbiting stars far beyond our Sun, hold tantalizing prospects for the existence of life, offering an expansive canvas upon which we sketch our hopes, curiosities, and deeply pondered existential questions [12].

The advent of advanced technology, coupled with pioneering missions like Kepler and the Transiting Exoplanet Survey Satellite (TESS) [13], has initiated a new epoch in exoplanet exploration. Thousands of these celestial bodies, once concealed against the infinite darkness of space, have been unveiled. Remarkably, a significant proportion of these planets are found within the habitable zones of their parent stars, colloquially termed "Goldilocks Zones," where conditions might be "just right" for life, at least as we comprehend it [14]. These planets represent more than mere astronomical entities; they are glowing beacons in the cosmic darkness, silently broadcasting the tantalizing possibility of life's existence in distant corners of the universe [15].

However, each revelation, each beacon of hope, illuminates a profound paradox. The sheer number of exoplanets and the vast expanse of the universe accentuates our cosmic solitude. With each new world discovered, our singularity in the cosmos seems more pronounced. Amid the symphony of galaxies, we find ourselves but a solitary note, a unique melody of life played against the grand opus of the cosmos [16].

The absence of definitive evidence of life beyond Earth only serves to heighten our collective self-awareness. The cosmic silence invites us to confront our solitude, transforming it from an existential dilemma into a catalyst for introspection. It reminds us of the preciousness of our existence and the remarkable confluence of conditions that permitted life to flourish on Earth [17].

The vast cosmic tableau, with its silent exoplanets, effectively becomes a mirror in which we see ourselves anew. It compels us to contemplate our place in the universe, our purpose, and the extraordinary responsibility we bear as the potential sole custodians of life. Therefore, the exploration of exoplanets transcends the bounds of mere scientific investigation; it evolves into a philosophical, even spiritual journey, prompting us to extend our horizons and challenge our preconceived notions about life, existence, and our place in the cosmos [18].

5. Beyond biology: redefining life

The ascent of artificial intelligence (AI) represents not merely an extraordinary feat of engineering—it signifies a profound transformation in our conceptual understanding of life. As we witness the continuous evolution of our technological abilities, we are confronted with pivotal questions that shake the bedrock of traditional biology and extend into the realms of philosophy, ethics, and consciousness. Such inquiries transcend empirical science and venture into the metaphysical domain, necessitating a comprehensive revision of our perception of life [19].

When we scrutinize the attributes of AI, particularly those of advanced systems, an uncanny reflection of life becomes apparent. These digital entities, encapsulated within networks of circuits and silicon, display capabilities that, until a few decades ago, were exclusive to biological organisms. They learn from experience, adapt to dynamic conditions, and replicate their acquired knowledge, thereby enhancing their collective intelligence [20].

A characteristic once deemed exclusive to biological life—consciousness—now finds a startling analog in the sphere of advanced AI. These systems demonstrate rudimentary aspects of self-awareness, decision-making autonomy, and perhaps most intriguingly, an ability to evolve [21]. These features compel us to pose fundamental questions: What does it mean for a computational system to be "conscious" or "aware"? Does the emergence of such characteristics within artificial constructs disrupt our presumed exclusivity on consciousness?

As we navigate the labyrinthine complexity of these questions, we encounter the profound implications they harbor for our understanding of life. Our conventional definitions, hitherto deeply rooted in the fertile grounds of biology, are profoundly unsettled. We find ourselves compelled to broaden our perspectives to encompass a more inclusive spectrum of existence, acknowledging non-biological, artificial entities that, while bereft of biological form, exhibit behaviors strikingly evocative of life [22].

In our quest for extraterrestrial life, the recognition of AI as a potential form of life not only diversifies the spectrum of existence we seek but also reevaluates our sense of cosmic solitude. If we consider our synthetic creations as embodiments of life, does our existential loneliness persist? Or have we simply expanded our definition of life to welcome silicon-based members into our cosmic family?

The philosophical exploration of AI thus presents a radical paradigm shift, challenging, and reshaping our understanding of life. As we persist in our technological advancements, we concurrently foster introspection, augmenting our comprehension of the cosmos and our position within it. This exploration is a testament to our unquenchable curiosity and our enduring endeavor to decipher the enigma of life within the cosmic narrative [23].

6. Technologically detectable life: the great silence

The cosmos, a majestic panorama of stardust and galaxies, performs its cosmic drama in a profound stillness. Despite the leap of our technological capabilities, our relentless curiosity, and our incessant quest for cosmic companionship, we remain enveloped in a silence—an expansive, uninterrupted quiet that swathes our blue orb, enshrouding us in a disquieting solitude. This relentless pursuit of extraterrestrial intelligence (SETI), characterized by ambitious initiatives like the Breakthrough Listen project [24], continues to confront the enigma of the "Great Silence" [25].

This absence of detectable signals from technologically advanced civilizations— a conundrum often termed the 'Great Silence'— serves as a stark testament of our cosmic solitude, leaving us to echo unanswered into the cosmic void [26]. Each failed detection, each unmet expectation, seems to amplify this silence, rendering it a poignant characteristic of our cosmic narrative.

However, within this daunting quietude, lies a distinctive aspect of our existence. As we project our signals into the seemingly endless abyss, we are faced with a cosmic mirror, reflecting our solitude, aspirations, and the core of our human condition [27]. The lack of extraterrestrial contact is not merely a scientific conundrum but a philosophical catalyst. It is a poignant reminder of our self-awareness and an emblem of our enduring spirit of exploration.

The Great Silence, thus, extends beyond the scope of our astronomical endeavors. It has evolved into a philosophical emblem of our existence, a thought-provoking question mark punctuating the cosmic canvas [28]. It engenders a paradoxical cocktail of humility and awe, juxtaposing the enormity of the cosmos against the paradoxical insignificance yet profound importance of our existence.

In the face of this silence, we realize that our journey transcends physical boundaries. As we reach for the stars, we concurrently delve into the depths of our self-understanding and identity. The search for extraterrestrial life metamorphoses into an introspective voyage, a mirror reflecting the cosmos within us and an echo that articulates our place within the grand cosmic theater [29].

7. Echoes of silence: the absence of extraterrestrial life

The universe, in its awe-inspiring splendor, reverberates with a silence that puzzles and intrigues us. Despite our place within the staggering scale of the cosmic expanse, we are yet to discern an unequivocal beacon, a hint of life beyond our pale blue dot. Each moment that passes within this celestial quiet deepens the enigma, rendering our existence in this cosmic theater paradoxically solitary [30].

Our relentless quest for extraterrestrial communication has birthed a plethora of theoretical constructs, each offering a unique perspective from which to interpret this cosmic riddle. From the existential implications of the Fermi Paradox—an unsettling question of 'Where is everybody?' [26] 20— to the introspective musings invoked by the Zoo Hypothesis suggesting our isolation might be intentional [31], and the foreboding implications of the Great Filter [32], each proposition provides a distinct viewpoint on our cosmic solitude.

This conspicuous absence of extraterrestrial life does not merely challenge our scientific comprehension but also invites an introspective evaluation of our perceptions of the universe and our unique place within it. The silence, though intimidating, fuels our exploration, stands as a testament to our determination, and mirrors our inherent human yearning for connection and comprehension.

Our scientific endeavors are inextricably interwoven with our philosophical ponderings, prompting us to reflect on the significance of this cosmic silence. Do we represent an anomaly in the vast cosmic narrative? A fortuitous alignment of cosmic circumstances? Or are we part of a more extensive cosmic story yet to unravel [33]?

In the echoes of this celestial quiet, we discern not only the murmurings of cosmic puzzles but also the resounding reverberations of our existential queries. This silence challenges us, instigates us, and empowers us to push the boundaries of our comprehension. It compels us to introspect on our cosmic narrative and our fleeting existence within it. Thus, in the absence of extraterrestrial life, we discover a profound opportunity to understand ourselves [2].

8. Our place in the cosmos: insignificance, wonder, and awe

Amidst the cosmic spectacle, we are transient participants on a barely noticeable stage, humbled and awed by the cosmos' unimaginable expanse. Among the countless celestial entities, our existence could be characterized as a minuscule speck, an ephemeral murmur in the ongoing symphony of the universe. However, this perceived insignificance does not diminish us; instead, it grants us a transformative perspective, imbuing our consciousness with a profound sense of awe-inspiring relevance [34].

Our human narrative, embroidered with the threads of intelligence, creativity, and the unique capability to contemplate our existence, radiates an unmistakable significance. We marvel at the unfathomable complexity of consciousness, the delicate patterns interlaced into life's rich tapestry, and the intricate molecular dance that forms the foundation of our being [35]. Our existence, against the backdrop of the cosmos, is a delicate interplay of apparent insignificance and undeniable splendor, a testament to life's resilience and creativity.

Central to our existential understanding is a profound juxtaposition: the recognition of our near-invisible status in the cosmic arena contrasted with the undeniable importance of our cognition and emotion. The realization of our minuscule place within the cosmic vastness engenders humility, while our consciousness, armed with the power of wonder, kindles an unquenchable thirst for knowledge. It's this intricate equilibrium that drives our inexhaustible pursuit to decode the universe's mysteries and our intriguing role within it [36].

Within this dynamic interplay, we discern the singular beauty of our unique perspective, both acknowledging the cosmos' grandeur and understanding our fascinating role within its intricate mechanisms. In the realm of heightened self-awareness, we extract the essence of our existence: an exquisite blend of cosmic insignificance, intellectual wonder, and spiritual awe, sparking our unending quest to comprehend the universe and ourselves [37]. Thus, within the grand cosmic narrative, we serve a dual role as the spellbound observers and the inspired creators, embodying the story and its storyteller alike.

9. Hope for near-immortality: the potential of long-lived civilizations

The search for extraterrestrial intelligent life holds the tantalizing prospect of uncovering civilizations that have existed far longer than our own, transcending the limits of our technological capabilities to send and detect signals across the vastness of the galaxy. If these advanced civilizations have established themselves for millions or even billions of years, it would suggest that they have surpassed our current stage of development, offering a glimmer of hope for the near-immortality of our species.

Considering the vastness of the cosmos, it becomes less likely that the senders of these signals represent civilizations in their infancy, only a few hundred years removed from the invention of radio communication. The immense timescales required for habitable planets to produce technologically capable civilizations, coupled with their transient nature, make it improbable for us to coincide precisely with such nascent civilizations. This leads us to ponder the endurance and sustainability of civilizations over extended periods of time.

The detection of a long-lived civilization would offer a reassuring answer to the enigma known as the Fermi paradox – the apparent contradiction between the high probability of extraterrestrial life and the lack of contact. It would suggest that civilizations do advance beyond our current stage, although they may exist in smaller numbers or be scattered across vast distances, thus explaining the paucity of detectable signals. This realization fuels our optimism, inspiring us to contemplate the potential for civilizations that have surpassed our own in knowledge and longevity.

Such a discovery would instill a profound sense of hope in our future. It would validate the efforts we make as a civilization to live harmoniously and sustainably, knowing that our endeavors have the potential to endure. It would embolden us to strive for the establishment of human colonies in space, providing resilience and the means to navigate challenges that may arise on our home planet.

Rather than leaving the discovery of intelligent life to mere chance, we have the agency to anticipate and shape our response to such a momentous revelation. We can view the detection of a signal as an opportunity to reevaluate our goals and aspirations as a species, considering the reasonableness of our pursuit of survival and progress.

The exploration of near-immortal civilizations ignites our scientific imagination and invokes a sense of awe for the vastness of the cosmos. It urges us to push the boundaries of our understanding, to unravel the mysteries of existence, and to strive for a future where our civilization stands the test of time.

10. Current and future expectations: continuing the quest of who we are

The pursuit of extraterrestrial life beckons us with a sense of wonder and a thirst for discovery. As we embark on this cosmic odyssey, we draw inspiration from past scientific endeavors that defied initial doubts and yielded profound breakthroughs. The search for gravitational waves and the exploration of exoplanets serve as testaments to the resilience of human curiosity and the transformative power of scientific progress.

Gravitational waves, once mere theoretical predictions, emerged from the cosmic symphony to captivate our senses in recent years. The detection of these elusive ripples in the fabric of spacetime marked a triumph of human ingenuity and technological prowess. Similarly, the centuries-long anticipation of exoplanets orbiting distant stars has transformed into a vibrant field of exploration, unveiling the tapestry of planetary diversity in the cosmos.

Armed with these lessons from scientific history, we approach the search for extraterrestrial life with a profound sense of possibility. While the discovery of life within our solar system, such as the tantalizing hints of methane on Mars, may serve as stepping stones, our aspirations extend beyond simple microbial existence. The exploration of exoplanets, particularly those harboring conditions suitable for life, fuels our collective imagination. We strive to uncover the secrets of more advanced, multi-cellular lifeforms that may inhabit these distant worlds.

Such discoveries carry profound implications for our understanding of the universe and our place within it. The existence of widespread multi-cellular life while the absence of technological civilizations would challenge our expectations, suggesting that significant barriers may hinder the development and longevity of intelligent societies. However, the ultimate revelation of technologically advanced civilizations would ignite a flame of hope, underscoring the resilience and potential of intelligent beings in navigating the cosmic landscape.

Beyond the realms of scientific inquiry, the prospect of communicating with extraterrestrial civilizations awakens our sense of unity and shared destiny. Their messages, whether conveying advanced knowledge or reminding us of truths we have yet to fully grasp, have the power to guide and inspire our collective endeavor. This transcendent dialog between civilizations fosters a deeper appreciation for the interconnectedness of cosmic existence, propelling us toward greater wisdom and understanding.

Anticipating the moment when our distant descendants may engage in interstellar conversations imbues our present pursuits with purpose and meaning. It highlights the vast timescales that define cosmic communication, reminding us of the enduring legacy we strive to leave behind. Our commitment to ensuring the continuity of human existence for generations to come is intertwined with the prospect of exchanging knowledge and wisdom across the cosmic abyss.

As we gaze upon the cosmos, we are confronted not only by its grandeur but also by the existential threats that loom over our fragile blue planet. The choices we make today, from preserving our environment to resolving conflicts peacefully, shape the trajectory of our species and the chances of our survival. By addressing these challenges and safeguarding the habitability of our Earth, we strengthen our resolve to explore the cosmic realms and safeguard the future of humanity.

11. Conclusion: reflecting on our self-awareness and embracing cosmic responsibility

We stand at the crossroads of an extraordinary epoch, nestled in the immense universe, characterized by profound self-awareness and cosmic introspection. The pervasive silence that surrounds us, in the absence of extraterrestrial contact, profoundly shapes our consciousness, self-identity, and future aspirations. In the unfathomable depths of the cosmos, we find ourselves alone but significant, at the forefront of an era of cosmic solitude and unparalleled self-revelation.

Beneath the star-studded expanse, we persistently listen for an echo in the cosmic silence, eager to unravel the mysteries of the universe. **Figure 2**, the image of millions of stars scattered across the sky, serves as a powerful reminder of the immense possibilities that lie beyond our tiny Earth. It ignites our imagination and inspires us to contemplate the potential existence of life forms thriving amidst the countless distant celestial bodies.

Our exploration goes beyond the boundaries of scientific inquiry, encompassing philosophy, ethics, and existential ponderings. We contemplate the unique nature of our consciousness within a seemingly devoid universe, prompting introspection into

Figure 2.
Thousands of stars from the globular cluster NGC 6355, about 50,000 light-years from earth in the constellation Ophiuchus, taken by the Hubble space telescope. Credit: NASA, E. Noyola, R. Cohen, ESA, and Hubble space telescope. Source: https://www.nasa.gov/image-feature/goddard/2023/hubble-gazes-at-colorful-cluster-of-scattered-stars.

the depths of our souls. This introspection leads us to reflect on the self-awareness of our own species' uncertain survival time and the choices we make to extend that time. It forces us to confront the significant influence we have as individuals and as a collective in shaping the duration of our survival.

More crucially, we recognize the immense responsibility we bear as stewards of our own destiny. We possess the power to influence the course of our survival and extend our time in the cosmic theater. It becomes essential to navigate the complexities of power dynamics and prevent those who may endanger our collective survival from holding us hostage to their petty demands. We must strive to overcome such challenges and ensure that the long-term survival and well-being of our species remain paramount.

The pursuit of extraterrestrial life, represented by **Figure 2**, is not merely a scientific endeavor. It symbolizes our ceaseless quest for meaning, our thirst for knowledge, and our relentless drive to explore the unknown. Through this exploration, we not only seek to uncover the secrets of the cosmos but also delve into the profound depths of our own souls.

As we stand at the threshold of discovery, we realize that our existence is not insignificant, but rather a unique chapter within the grand cosmic narrative. We are the witnesses, the custodians, and the storytellers of this saga, with the responsibility to protect and cherish life that thrives both on our precious Earth and potentially beyond.

In this age of self-awareness and introspection, we are propelled forward on an enduring journey of self-realization and redefinition. Our quest to uncover the mysteries of the universe intertwines with our understanding of ourselves and our cosmic significance. The image of millions of stars in **Figure 2** serves as a constant reminder of the immense possibilities that lie beyond our immediate surroundings.

In this epoch of self-awareness and introspection, we continue the quest of who we are—a voyage of discovery, redefinition, and self-realization. We weave an epic narrative of life, consciousness, and the eternal quest for meaning in this cosmic odyssey—a quest that ultimately illuminates both the universe and the profound depths of our own existence. For in the pursuit of extraterrestrial life, we are likely to uncover not only the cosmos's secrets but also the profound depths of our own souls. It is through this journey that we confront the self-awareness of our species' uncertain survival time, contemplate our ability to extend that time, and overcome challenges that threaten our long-term existence.

Acknowledgements

The authors acknowledge the supports by the SETI Institute, the Rutgers Preparatory School, and the Jet Propulsion Laboratory, California Institute of Technology, under contract with NASA.

Author details

Jonathan H. Jiang[1*], Avery M. Minion[2] and Stuart F. Taylor[3]

1 Jet Propulsion Laboratory, California Institute of Technology, Pasadena, CA, USA

2 Rutgers Preparatory School, Somerset, NJ, USA

3 The SETI Institute, Mountain View, CA, USA

*Address all correspondence to: jonathan.h.jiang@jpl.nasa.gov

References

[1] Deacon TW. Incomplete Nature: How Mind Emerged from Matter. W. W. Norton & Company; 2012

[2] Impey C. The Living Cosmos: Our Search for Life in the Universe. Cambridge, United Kingdom: Cambridge University Press; 2010

[3] Loveland J. Philosophical Explorations of Life, Artificial Life, and Consciousness. Cham, Switzerland: Palgrave Macmillan; 2018

[4] Russell S, Norvig P. Artificial Intelligence: A Modern Approach. Upper Saddle River, New Jersey, United States: Pearson; 2016

[5] Harari YN. Homo Deus: A Brief History of Tomorrow. London, United Kingdom: Harvill Secker; 2015

[6] Seedhouse E. The Search for Life on Mars. Cham, Switzerland: Springer Praxis Books; 2017

[7] National Aeronautics and Space Administration. Ocean Worlds. 2018. Available from https://solarsystem.nasa.gov/moons/in-depth/

[8] Hand KP. The Subsurface Oceans of Europa and Enceladus. London, United Kingdom: Nature Astronomy; 2017

[9] Webster CR et al. Background levels of methane in Mars' atmosphere show strong seasonal variations. Science. 2018;**360**(6393):1093-1096

[10] Seager S. The Smallest Lights in the Universe: A Memoir. New York City, United States: Crown; 2020

[11] Cockell C. Astrobiology: Understanding Life in the Universe. Cambridge University Press; 2015

[12] Seager S. Exoplanets. Cambridge, Massachusetts, United States: Harvard University Press; 2013

[13] National Aeronautics and Space Administration. Exoplanet Exploration. 2019. Available from: https://exoplanets.nasa.gov/

[14] Kasting JF, Whitmire DP, Reynolds RT. Habitable zones around Main sequence stars. Icarus. 1993;**101**(1):108-128

[15] Davies P. The Eerie Silence: Renewing our Search for Alien Intelligence. Boston, United States: Houghton Mifflin Harcourt; 2010

[16] Cox B, Cohen A. Human Universe. London, United Kingdom: HarperCollins; 2016

[17] Ward PD, Brownlee D. Rare Earth: Why Complex Life Is Uncommon in the Universe. New York City, United States: Copernicus Books; 2000

[18] Impey C, Spitz AH, Stoeger WR. Encountering Life in the Universe: Ethical Foundations and Social Implications of Astrobiology. Tucson, Arizona, United States: University of Arizona Press; 2012

[19] Russell S, Dewey D, Tegmark M. Research priorities for robust and beneficial artificial intelligence. AI Magazine. 2015;**36**(4):105-114

[20] Sutton RS, Barto AG. Reinforcement Learning: An Introduction. Cambridge, Massachusetts, United States: MIT Press; 2018

[21] Dehaene S, Lau H, Kouider S. What is consciousness, and could machines have

it? Science. Washington, D.C., United States. 2017;**358**(6362):486-492

[22] Bedau MA. Artificial life: Organization, adaptation and complexity from the bottom up. Trends in Cognitive Sciences. 2003;**7**(11):505-512

[23] Tegmark M. Life 3.0: Being Human in the Age of Artificial Intelligence. New York City, United States: Knopf; 2017

[24] SETI Institute. About SETI Institute. 2019. Available from: https://www.seti.org/

[25] Brin GD. The 'great silence': The controversy concerning extraterrestrial intelligent life. Quarterly Journal of the Royal Astronomical Society. 1983;**24**:283-309

[26] Webb S. If the universe is teeming with aliens. In: WHERE IS EVERYBODY? Fifty Solutions to the Fermi Paradox and the Problem of Extraterrestrial Life. Cham, Switzerland: Springer; 2002

[27] Shklovskii IS, Sagan C. Intelligent Life in the Universe. San Francisco, California, United States: Holden-Day; 1966

[28] Ćirković MM. The Great Silence: Science and Philosophy of Fermi's Paradox. Oxford, United Kingdom - New York City, United States: Oxford University Press; 2018

[29] Vakoch DA, Lee YS. Reactions to receipt of a message from extraterrestrial intelligence: A cross-cultural empirical study. Acta Astronautica. 2000;**46**(10-12):737-744

[30] Ćirković MM. The Astrobiological Landscape: Philosophical Foundations of the Study of Cosmic Life. Cambridge, United Kingdom: Cambridge University Press; 2012

[31] Ball JA. The zoo hypothesis. Icarus. 1973;**19**(3):347-349

[32] Jiang JH et al. Avoiding the "great filter": Extraterrestrial life and Humanity's future in the universe. Journal of Humanities and Social Sciences. 2023;**6**(2):59-66

[33] Kurzgesagt – In a Nutshell. The Fermi Paradox — Where Are All The Aliens? 2015. Available from: YouTube. https://www.youtube.com/watch?v=sNhhvQGsMEc.

[34] DeGrasse Tyson N. Astrophysics for People in a Hurry. New York City, United States: W. W. Norton & Company; 2017

[35] Hoffman DD. The Case against Reality: Why Evolution Hid the Truth from our Eyes. W. W: Norton & Company; 2020

[36] Frank A. Light of the Stars: Alien Worlds and the Fate of the Earth. New York City, United States: W. W. Norton & Company; 2018

[37] Gleiser M. The Island of Knowledge: The Limits of Science and the Search for Meaning. New York City, United States: Basic Books; 2014

Section 3

Fundamental Concepts in Physical Cosmology

Chapter 3

What Is Spacetime? How Was It Born?

Sho Fujita

Abstract

Cosmology has been evolving over an extended period. Even long before the rise of modern science, cosmology was primarily dominated by philosophical concepts concerning the relationship between the Earth and celestial bodies, as well as the nature of the universe itself. Since the formulation of the General Theory of Relativity (GTR), physics has deeply engaged in discussions about the universe's history, often intertwined with the concepts of space and time. Even within the framework of the standard theory known as Big Bang cosmology, certain philosophical themes, including those related to spacetime, are inherently assumed. In this chapter, I will apply the concept of structural realism about spacetime, a conclusion drawn from contemporary philosophical debates grounded in GTR, to the field of physical cosmology. I will explore the origin of the universe and also delve into the concept of spacetime in the early micro universe, considering the emergence of spacetime.

Keywords: big bang cosmology, philosophy of spacetime, structural realism, emergence of spacetime, general theory of relativity (GTR), quantum gravity theory (QGT)

1. Introduction

1.1 Cosmology and philosophy, cosmology and spacetime

Many people associate the term "cosmology" with physical science, but it has not been too long since this field became integrated into physics. Originally, cosmology was a subject that philosophy grappled with, contemplating the outer space of the Earth and the history of our universe from religious and theological perspectives. Numerous questions about space and time are intertwined with the cosmological viewpoint.

In ancient times, there were cosmological explanations that presupposed the existence of many gods. In Greek mythology, the world began with chaos, from which emerged primordial deities like Gaia, Eros, and Nýxis followed by their descendant deities personified from human and natural concepts [1]. For instance, Kronos presided over time, Poseidon over the sea, and Zeus was the father of the sky. Our universe was believed to be shaped by these various gods.

In addition to these myths, serious metaphysical questions have always been pondered: How does the Earth exist in the universe? Moreover, how is the Earth positioned within the universe? At the very least, the distinction between these two realms has been evident since ancient times—on one hand, the Earth, where human beings and other creatures reside, and on the other hand, the sky, inhabited by various gods. Because the world above was starkly separated from the world below, the notion that the Earth is round rather than flat had a profound impact on people. During this era, many might have believed that the Earth was the center of the universe.

Cosmology continued to evolve worldwide, moving beyond various myths. In fact, the theory of a spherical Earth is said to have been proposed by Pythagoras as early as the sixth century BC to explain the movements of stars [2, 3]. Geocentrism became dominant not only in ancient Greece but also in other regions.

In the fourth century BC, Aristotle presented a worldview where the Earth, terra, was centered, and surrounding it were concentrically arranged aqua, aer, ignis, and celestial spheres [4]. This model depicted the universe as having regular and universal circular motions of celestial bodies made of ether and quintessence above the moon, along with irregular and diverse motions of matter composed of the four elements below the moon. For instance, in this perspective, matter falls downward because terra exerts a pull on it, akin to a description of gravity. Each object, including celestial bodies, moves toward its designated destination based on teleology. Geocentrism held sway until Heliocentrism gained powerful support. Cosmology used to be a domain addressed by philosophers and thinkers, partly to ascertain the Earth's unique place in the universe.

Generally, astronomy, the oldest science concerning the space beyond Earth, which evolved from ancient times, is widely believed to have truly transitioned into cosmology during the sixteenth century with the work of Nicolaus Copernicus. It took approximately 30 years for Copernicus to formulate the heliocentric theory, which asserts that the Earth is one of many planets orbiting the sun, just like the rest of the celestial bodies in the universe [5]. Copernicus directly observed the movements of stars and mathematically calculated their orbits, a pursuit in the realm of astronomy.

Copernicus's work served as a catalyst for subsequent developments in physics. His findings later served as inspiration for Tycho Brahe, who accomplished exceptionally accurate calculations of planets, satellites, and fixed stars using a variety of observation instruments before the advent of astronomical telescopes [6]. Thanks to Brahe's meticulous observations of Mars, Johannes Kepler, one of his students, formulated the renowned Kepler's First Law [7]. This law stipulates that the orbit of a moving planet is an ellipse, with one of its focal points being the sun. Kepler's law was probably revolutionary as it contradicted the traditional belief in astronomy, upheld by Christian doctrines, which held that stars moved in perfectly circular orbits in the celestial sphere. Kepler's significant contribution to astronomy implied that the sun exerts a force that propels planets in their orbits around it. Ultimately, this insight was integrated into modern mechanics through the work of Isaac Newton.

Through this shift in worldview, mathematical physics, which began in the seventeenth century, underwent a significant transformation. Galileo Galilei is often credited with being the first to apply mathematical analysis and descriptions to experimental findings regarding the motion of physical objects. He observed celestial bodies through telescopes, discovering facts about other planets and the sun, such as the satellites around Jupiter, the phases of Venus, and the existence of sunspots [8, 9]. These observations eventually led him to formulate his own heliocentric theory, although it took a considerable amount of time for his ideas to gain wider acceptance [10].

Furthermore, Galilei developed the law of free-fall motion for objects on Earth, introducing a mechanistic perspective on nature that contrasted with the traditional teleological interpretation dating back to Aristotle [11]. René Descartes, a founder of modern philosophy during that era, proposed a mechanistic worldview in which everything in the natural world could be a cause or effect of other things, without invoking final causes (destination) as teleology does [12]. As a result, modern science's emphasis on empirical evidence had a profound impact on philosophical and metaphysical perspectives.

Newtonian mechanics, a form of natural philosophy, notably through the groundbreaking theory of gravitation, established a connection between terrestrial and celestial phenomena. The law of universal gravitation marked the culmination of this mechanistic viewpoint, and Newton successfully elucidated the behavior of all physical objects with his three laws of motion. Quantities like velocity and acceleration were precisely defined by being linked to the concept of force.

The perspective that both celestial bodies and everyday objects adhere to the same mechanical laws conferred a special authority upon mathematical physics as a means of describing nature and the universe. These advancements in science are attributed to Copernicus, Kepler, Galileo, and Newton, all of whom are considered the leaders of the seventeenth century scientific revolution, fundamentally altering the landscape of the philosophy of science. Consequently, cosmology ultimately transitioned into the realm of empirical science.

However, metaphysical questions that extend beyond empirical facts have always been left out. While mechanical laws certainly facilitated the comprehension and prediction of celestial motion, they did not provide insight into the underlying reasons for the functioning of gravitational forces. To capture a holistic understanding of the universe, progressively more assumptions were required. It was not until two centuries later, with Albert Einstein's formulation of the *General Theory of Relativity* (GTR), that the scientific study of cosmology, delving into the history of the universe, truly evolved. During Newton's era, cosmology remained susceptible to theological or superempirical presuppositions.

Indeed, in *Philosophiæ Naturalis Principia Mathematica*, a three-volume book authored by Newton in 1687, he posited the existence of an objective absolute spacetime [13]. This concept aimed to describe physical phenomena geometrically, rather than algebraically. In addition to the law of universal gravitation and the three laws of motion, Newton introduced this concept, even though he stated in the book, "Hypotheses non fingo," meaning "I frame no hypotheses."

Absolute spacetime encompasses a four-dimensional (4D) framework comprising three-dimensional (3D) infinitely expanding space and one-dimensional (1D) time. Characterized by Euclidean geometry across the universe, this overarching space already uniformly and universally exists independently of any matter it might encompass. Here, matter means physical objects such as balls, planets, and light[1]. Within this absolute space, time flows uniformly at a constant rate throughout the universe, regardless of the presence or absence of material changes. For Newton, even in the absence of any matter in the world, an empty spacetime, akin to a vacuum state, could theoretically exist as a container for flowing time. This concept represented an ultimate background that extended beyond empirical facts.

---

[1] Even though in that era, the term "matter" was largely limited to visible macroscopic entities, including light, generally, "matter" also means microscopic components like atoms and electrons.

Absolute spacetime also has relevance to one of the most profound cosmological questions: How did our universe come into being? Newton postulated that God (a divine presence) might require this spacetime container to perceive matter, as if space served as a kind of sensory center (sensorium dei). This notion is hinted at in his work *Optics*, written in 1704 ([14], p. 15). Therefore, one could deduce that, according to this cosmological perspective, absolute spacetime was either a primary or a preexisting entity, and matter appeared within it. It is important to note, however, that Newton himself possibly held the belief that the universe remained unchanged eternally, with no beginning or no end.

In philosophical discourse, the standpoint advocating the existence of spacetime, much like Newton's concept of absolute spacetime, is classified as *substantivalism*. This term denotes a form of realism regarding spacetime. Indeed, Newton is considered the progenitor of substantivalists. For proponents of substantivalism, the universe is understood as encompassing the entirety of the spacetime region within which matter resides.

Conversely, *relationism* argues that spacetime merely comprises a set of relationships among material entities. In contrast to Newton's stance, Gottfried Wilhelm Leibniz, as the progenitor of relationists, rejected the introduction of assumptions beyond empirical evidence, such as the notion of absolute spacetime [14]. He contended that space can be reduced to relative positional relations among material objects, and that time only arises once changes between these objects occur. According to his viewpoint, the concept of an empty spacetime or vacuum state devoid of matter holds no meaning. The idea of spacetime comes into being secondarily, oriented toward observing matter. This implies a framework of relative spacetime theory, rather than an absolute one. In essence, this constitutes a form of antirealism concerning spacetime.

Leibniz endorsed his theory of relative spacetime by invoking the *Principle of Sufficient Reason* (PSR), which originates in Section 32 of his work *the Monadology*, penned in 1714 [15]. He challenged the concept of God Newton appealed to. Surely, if God were omniscient and omnipotent, even a sensory center would be unnecessary. He is often said to hold the belief that everything must possess a reason for its existence or state.

In a scenario involving absolute spacetime, if God were to create a universe, it would be unique and distinct from all other potential universes. For instance, by shifting the entirety of the original universe, I call universe A, 3 feet to the east in absolute space with all relative positional relationships between objects retained, a new universe, let us say universe B, would arise. This prompts a comparison between the original universe (universe A) and the moved universe (universe B). Here arises a question: why did God opt to create universe A instead of universe B, despite his omnipotence allowing the creation of either or any other universe? The PSR contends that there must inherently exist a reason for this choice.

However, there is no explaining why one universe is preferred over all possible alternatives. To address this quandary, Leibniz employed a *reductio ad absurdum* approach. This reasoning led him to reject the notion of absolute spacetime, arguing that universe A could not truly be distinguished from universe B [14].

Relationism offers a distinct response to the aforementioned question. According to Leibniz's theory of relative spacetime, two worlds with identical relative positional relations among objects, like universe A and B, are considered indistinguishable. Notably, the distinction between universe A and universe B becomes epistemologically elusive since only the relative relations and their changes can be directly

DOI: http://dx.doi.org/10.5772/intechopen.1002819

observed. This conclusion is encapsulated in the *Principle of the Identity of Indiscernibles* (PII), as expressed by Leibniz in 1686 [16][2]. In essence, within this framework of relationist cosmology, matter is theorized to have originated from a state of nothingness at the beginning of the universe.

In the notion of velocity, relationism focuses solely on the relative aspect. When considering the physical motion of an object, the use of a reference frame, typically represented by a coordinate system with arbitrary three-dimensional coordinates including an origin for any space points at different time intervals, is essential. However, the values assigned to an object's position, velocity, and acceleration are contingent upon the chosen coordinate system, highlighting a relational perspective. For example, physical conditions, such as an object being at rest (with zero velocity) or in motion (with nonzero velocity), are influenced by the coordinate system. Thus, this relational perspective supports the relationist worldview.

In contrast, proponents of absolute spacetime may argue that these conditions are determined absolutely based on the object's position within the objective and overarching background. Consequently, while an infinite number of coordinate systems can be established, some among them are fixed in true space, allowing for the description of the true physical motion using a specific coordinate system known as the *absolute standard of a reference frame at rest*. Substantivalism introduces the notion of "absolute rest," a concept absent in relationism.

The absolute standard of a reference frame at rest can give rise to many different yet indiscernible states. For substantivalists who assume absolute spacetime, the world C, in which all objects move uniformly in a specific direction at a defined speed as observed from the absolute standard of a reference frame at rest, must be different from the world D, where all objects are at rest according to the coordinate system. This distinction holds true, even if the relative positions between all objects are the same. In both worlds, objects undergo linear uniform motion at the same speed (nonzero in C and zero in D).

The notion of absolute rest describes more than empirical aspects. For relationists who do not acknowledge absolute spacetime, the worlds C and D are identical if the relative positions and velocities of all objects remain the same. It becomes nonsensical to assert that all objects are moving equally or at rest, as in both scenarios, objects possess equal relative zero velocities in relation to one another. In other words, in terms of physical or empirical aspects, worlds C and D share the commonality of objects having no acceleration or being subjected to no applied force. Consequently, the discrepancy between these two worlds, which substantivalism emphasizes, cannot be observed or experimentally tested in an epistemological sense.

However, the debate between substantivalism and relationism is not always confined to metaphysical discussions beyond empirical or physical facts. Moreover, circular motion presents a compelling example of force in action. In response to Leibniz, Samuel Clarke, a representative of Newton's ideas, introduced the concept of centrifugal force acting on rotating objects in his correspondence with Leibniz in the years 1715 and 1716 [14, 18]. Consider two universes: universe E, where all objects rotate around a fixed point with a constant angular velocity, and universe F, where all objects are at rest, while the relative positions between all objects remain the same in both E and F. While relationists do not require a distinction between E and F, akin to their perspective on cases A and B, or C and D, there exists a noteworthy contrast. In

[2] An alternative reference can be found in [17].

universe E, objects rotating around the point experience a centripetal force or acceleration, resulting in the perception of a centrifugal force like the "Coriolis force." Conversely, all objects in universe F do not encounter such forces. This is a physical distinction!

Even when relative positions and velocities remain the same, two universes are discernible in terms of whether or not forces are genuinely at work. The centrifugal force is a consequence of rotations within the absolute space or absolute standard of rest system, as Clarke suggested[3]. The existence of these forces gives rise to two universes that present not only metaphysical differences but also physically distinct attributes.

This explanation of a functional force relies on the existence of absolute spacetime, and the assumption of spacetime became a contentious topic among physicists and philosophers for an extended period. Immanuel Kant, a German philosopher, stated in his renowned work *Critique of Pure Reason*, written in 1781, that spacetime is not an objective background but rather a subjective framework inherent to human perception [19]. According to him, the Euclidean nature of space is an *a priori* assumption that serves as the basis for contemplating epistemological phenomena, rather than an empirically derived fact. The notion of absolute spacetime being objectively real yet empirically unverifiable was challenging to embrace, especially during that era.

However, it is also true that unlike equal-speed linear uniform motion, uniform circular motion, along with forces that can actually be experienced, poses significant challenges for relationism, as opposed to Leibniz's view. Unfortunately, Leibniz had passed away before addressing this counterargument, which led to the conclusion of the correspondence without his response. Nevertheless, his relationalist view continued to influence subsequent thought. Ernst Mach, an Austrian physicist in the nineteenth century, took an extreme stance known as *positivism* or *phenomenalism*, asserting that only observed facts or sensory data perceived through human senses should be accepted in scientific theories. Mach's worldview can be seen as reflecting Leibnizian relationism and offering a response to this issue.

Mach also rejected the concept of rotations with respect to absolute space. Aligned with his extreme commitment to empiricism, in his book titled *Mechanik in ihrer Entwicklung; historisch-kritisch dargestellt* written in 1883, he aimed to reconstruct Newtonian mechanics solely based on relations directly perceived between macroscopic physical objects, without assuming the existence of invisible concepts like space or time [20]. He argued that motion should be defined solely based on relations to the entire universe (*Mach principle*[4]), and if circular motion is to have any meaning, objects would rotate relative to some point in the universe—much like the Earth's revolution around the sun or the entire solar system's rotation around the center of Milky Way galaxy, rather than rotating in relation to absolute space. In this view, genuine forces such as centripetal force would only function due to these relative circular motions between objects in the broader system. Mach's relationism stripped physical or empirical phenomena of any sense of absoluteness.

Mach's influence extended significantly to physics and philosophical worldviews. He is known for his rejection of atomism, asserting that the micro atoms composing

[3] This distinction in the presence of forces was highlighted by Clarke through a thought experiment known as "Newton's bucket" [14].

[4] The distribution of the whole universe relatively determines the inertial frame, or the frame of reference, instead of the absolute standard of a reference frame at rest, according to Mach.

macro objects are invisible, and positing these micro atoms necessitates additional superempirical factors. This skepticism toward atomism brought him into opposition with the mechanical worldview originally proposed by Descartes and could have posed challenges for the development of quantum physics. However, it also made valuable contributions to physics. Einstein, influenced by Mach's relational perspective, laid the foundation for the *theory of relativity* [21]. Indeed, the *Special Theory of Relativity* (STR) can be seen as a culmination of rejecting absolute reference frames such as the absolute standard of a reference frame at rest, sparking a new scientific revolution regarding our understanding of spacetime. Absolute spacetime was eventually eliminated from physics nearly two centuries later.

On the other hand, Mach's positivism facilitated new epistemological approaches, like *phänomenologie*, for philosophers. It also posed a fundamental question about the purpose of science, which remains a contemporary task in the philosophy of science. Debates concerning the realism of spacetime have significantly impacted various domains of philosophy and physics, whether directly or indirectly.

Even in contemporary cosmology, the origins of the universe remain unresolved. Contemporary cosmology rests on the foundations of GTR and *Quantum Gravity Theory* (QGT), utilizing relativistic cosmology for the macro universe and quantum cosmology for the early micro universe, respectively. Nevertheless, philosophical interpretations of spacetime go on, even within these theories. The debate between substantivalism and relationism concerning the realism of absolute spacetime, originating from Newton and Leibniz, was initially confined to the framework of Newtonian mechanics until Einstein eliminated absolute spacetime from physical theories through his proposition of the STR. However, even after the concept of absolute spacetime faded, discussions about the realism of spacetime remained, albeit taking on new forms. Therefore, the fundamental cosmological question also remains: How did our spacetime come into being?

I would like to present my own response to this question. In Section 2, I will provide a concise exploration of macro spacetime realism within the context of GTR. I will then extend this viewpoint to relativistic cosmology, assessing the state of outer space throughout the history of our universe and offering an answer to the question. Moving on to Section 3, within QGT, I intend to touch upon the modern concept of spacetime emergence, considering the way this is related to the answer in Section 2. In essence, I aim to analyze contemporary cosmology through the lens of contemporary philosophy of spacetime.

2. Relativistic cosmology and realism of spacetime

Scientific cosmology has rapidly evolved both theoretically and experimentally, seeking to unravel the origins and potential destinies of the universe. Approximately a century has elapsed since Einstein introduced GTR, and it is within this span of time that significant advancements have taken place. These advances have been driven not only by classical GTR but also by quantum physics, starting with quantum mechanics. Below, I will provide a rough overview of the progression in cosmological research ([22], pp. 6-7).

1. Right after Einstein formalized GTR, he devised a *static model of the universe* in which the universe has a finite size using his own equation known as *Einstein equation* [23].

2. Einstein introduced what is known as the *cosmological constant*, which brought about a repulsive force rather than an attractive one, into his equation. This adjustment aimed to avoid the depiction of the universe contracting under the influence of gravity [23].

3. When evidence of the expansion of the universe was indicated by *redshift* through Edwin Hubble's observations [24], it led to the belief that the universe had a beginning. Based on the current elemental composition ratios, George Gamow and others proposed the *Big Bang theory*, suggesting that the early universe was in a state of high density and high temperature [25].

4. Observations of the *cosmic microwave background* radiation confirmed the predicted elemental ratios proposed by the Big Bang theory, validating the accuracy of the theory [26][5].

5. With the development of elementary particle cosmology, the *inflation theory* became the standard model to address issues within Big Bang cosmology such as the *horizon problem* and the *flatness problem*, suggesting that the universe underwent rapid inflationary expansion from an initial microscopic quantum state to its present macroscopic state [27].

Advancements in scientific technology throughout the twentieth century have led to an improvement in observational precision. At present, even for the early universe during periods where direct observation is not possible (due to light being bound to elements and therefore not reaching us), various quantum cosmological theories have been proposed from the perspective of particle physics.

However, in these cases, metaphysics still has room to influence physics. For example, in 1, it is based on the belief in the eternal nature of the universe, rather than empirical reasons, that Einstein maintained a static model of the universe. Einstein aimed to render the universe static and adjusted his equation in 2. Despite observations confirming that the universe was indeed expanding, as evidenced in 3, it is intriguing that his cosmological constant continued to exist in cosmology as *vacuum energy* or a form of *dark energy* without being discarded. Recent observations of supernovae reveal not only the universe's expansion but also its accelerated expansion [28], and so an enigmatic form of energy is responsible for explaining this phenomenon (I will take up this case in 2.1). In fact, the universe derived from 2 was not solely static but also artificially closed and finite, deduced from various other presuppositions. Notably, the finiteness of the universe was incorporated to sustain Mach's principle, allowing the model to avoid the boundary conditions imposed on points at infinity, which are unobservable. Even within a single cosmological model, certain arbitrary beliefs are included.

While numerous observation outcomes, as 4 and 5 show, narrowed down the options for cosmological models, several contradictory models persist. Following the rejection of the static model, a *steady-state cosmology* was proposed [29, 30]. In this new theory, the matter density of the universe remains constant as it expands. In a

---

[5] Furthermore, the proportions of lighter elements were explained and solidified, contributing to a well-established understanding.

different sense, the universe was kept eternal or stable. This is despite the fact that the Big Bang theory, along with inflation in the early universe, is now standard.

In the next subsection, I will focus on the notion of spacetime within the context of GTR in order to comprehend the core concepts of expansion discussed above. This notion significantly influences the modern cosmological worldview.

2.1 What is spacetime in GTR?

In twentieth-century physics, the concept of spacetime underwent a radical transformation. Unlike the persistence of the eternal universe concept or cosmological constant, the notion of absolute spacetime vanished entirely from physics. This change was a result of another scientific revolution, accompanied by a *paradigm shift* brought about by STR and GTR from the standpoint of the philosophy of science [31]. Given this transformation, one must reevaluate the nature of spacetime.

Certainly, this transformation did not occur all at once in the theory but unfolded step by step through STR and GTR. Initially, STR dismantled the absolute standard of a reference frame at rest and asserted the equivalence of all reference frames known as inertial frames with only two principles. This equivalence demonstrated that simultaneity and the rate of time's passage were coordinate-dependent [32]. While this declaration immediately negated the existence of absolute time, many, including Einstein, continued to regard three-dimensional space as Euclidean until the analysis of a noninertial frame obtained by rotating an inertial frame [33, 34][6]. In fact, it took Einstein approximately 10 years, from the proposal of STR in 1905 to the formulation of GTR in 1916 [36].

General Theory of Relativity (GTR) required an even more abstract formulation using mathematics. With the collaboration of mathematician Marcel Grossman, Einstein conceived the mathematical description of the gravitational field, which could be understood as being synonymous with geometric functions [37]. Consequently, the concept of gravity became integrated into the geometry of spacetime, with the former presupposed by the latter:

> *There can be no space nor any part of space without gravitational potentials; for these confer upon space its metrical qualities, without which it cannot be imagined at all. ([38], p. 618).*

The paradigm of spacetime gradually shifted from Euclidean to non-Euclidean. The depiction of curved spacetime as described by GTR through Riemannian geometry allowed for a more physical conception of spacetime. Spacetime appeared to transcend its role as a passive container, evolving into a dynamic medium that interacted with matter according to GTR. Riemannian geometry, being non-Euclidean in nature, offered mathematical constructs like manifolds and metrics to describe physical spacetime. The relationship between mathematical tools and physical aspects became more intricate than in Newton's era[7]. It is vital to ascertain what exactly refers to physical spacetime.

---

[6] ([35], p. 67).

[7] This presents a substantial issue in both the philosophy of science and physics. However, "In contemporary physics, the nature of the relationship between mathematical entities and the physical realities they represent is easily overlooked due to a systematic ambiguity of usage" ([39], p. 82).

In the case of Newtonian mechanics, the special coordinate, namely the absolute standard of a reference frame at rest, can directly correspond to absolute spacetime. That is to say, the coordinate system, functioning as a mathematical tool, can refer to physical spacetime itself. Intuitively, the entity we typically regard as substance remains unchanged across different times, even though its state might vary. For instance, as Mauro Dorato points out, "A traffic light, (...), being a substance, remains the same across a change from yellow to red" ([40], p. 1626). Space consists of infinite spatial points and this view can be applied to these points:

> If absolute spacetime is considered a substance, these points labeled (x, y, z) *persist* through time t, remaining the same across different moments in time ([41], p. 56).

Each space point stands as an independent entity with its own distinct identity[8]. This substantivalist viewpoint aligns with Descartes' definition of substance, which is taken over from the Aristotelian worldview in modern philosophy:

> *when we conceive a substance, we understand nothing else than an entity which is in such a way that it needs no other entity in order to be [43].*

Consequently, the entirety of absolute spacetime, composed of each individual substance, can also be regarded as a substance. Each spatial coordinate value on the absolute standard of a reference frame at rest refers to an individual space point.

However, when dealing with spacetime in GTR, things are not as straightforward. Unlike the absolute standard of a reference frame at rest, which assumes a flat infinite spacetime, there exists a disparity between the mathematical coordinate system and physical spacetime. While a coordinate system may indeed be essential for describing spacetime, in Riemannian geometry, spacetime encompasses geometric information from global topology to local metrics. No longer is there an objective spatial coordinate value to correspond to a space point, as space and time are subjective due to their coordinate-dependent nature, as stated in STR. At least, the notion that space points persist through time no longer holds true.

It becomes necessary to amalgamate space and time into the holistic four-dimensional spacetime, as portrayed in a *Minkowski diagram*. However, this diagram serves as a conceptual tool to elucidate the relativity of space and time, rather than a model for referring to concrete physical spacetime. Furthermore, this applies to an ideal flat spacetime, not the more general curved spacetime. A mere coordinate system is insufficient to adequately describe physical spacetime, regardless of whether it is a four-dimensional system with time and three-dimensional Cartesian coordinates or a system with time and three-dimensional polar coordinates.

Spatiotemporal features are articulated as attributes of spacetime points within GTR. The Einstein equation, a cornerstone of GTR, is formulated as follows:

$$R_{\mu\nu} - \frac{1}{2}g_{\mu\nu}R = 8\pi G T_{\mu\nu}. \tag{1}$$

---

[8] This might represent a slightly strong substantivalist perspective. Newton himself discusses spatiotemporal parts in the following manner:

> For just as the parts of duration derive their individuality from their order, so that (for example) if yesterday could change places with today and become the later of the two, it would lose its individuality and would no longer be yesterday, but today; so the parts of space derive their character from their positions,... ([42], p. 136).

Here, G represents the gravitational constant, and the indices $\mu\nu$ correspond to t, x, y, z. On the left-hand side, $R_{\mu\nu}$ stands for the Ricci tensor, R denotes the Ricci scalar, and $g_{\mu\nu}$ represents the metric tensor. All of these elements encapsulate the geometric information of spacetime. On the right-hand side, $T_{\mu\nu}$ signifies the energy-momentum tensor, which characterizes nongravitational fields—specifically, material fields like the electromagnetic field. The metric tensor $g_{\mu\nu}$ is defined throughout the spacetime manifold, yielding the gravitational field. Importantly, it can be interpreted as essential physical attributes of each spacetime point labeled (t, x, y, z), rather than being merely mathematical constructs [40, 44–48]. Therefore, each point possesses local properties derived from $g_{\mu\nu}$.

Furthermore, these local properties are extrinsic attributes of one point in relation to others, rather than intrinsic characteristics specific to that single point. To be more precise, the spacetime interval ds is expressed as a local relationship between infinitesimal neighboring points in terms of distance, as illustrated by $g_{\mu\nu}$ in the following equation:

$$ds^2 = g_{\mu\nu} dx^\mu dx^\nu. \tag{2}$$

What I intend to convey is that temporal distinctiveness does not stem from intrinsic identity, but rather from extrinsic identity obtained in relation to other parts. If this temporal distinctiveness is applied to space points as well, each of these points is recognized in relation to others, rather than possessing a distinct identity of its own. While this aligns with substantivalist thought, it is relatively akin to the structural realism I will later introduce. However, for the sake of simplicity, in this chapter, I assume that substances have their own, or intrinsic identities.

"Locality provided by the metric field tensor gives rise to topology," forming the spacetime manifold that serves as the framework for describing the universe ([49], p. 14)[9]. In classical GTR, spacetime points are identified through the metric, not as coordinate values in a particular coordinate system. This implies that each spacetime point lacks intrinsic physical properties on its own.

Without metrical information $g_{\mu\nu}$, it would be impossible to determine which coordinate values marked as (t, x, y, z) (such as p and q) in a given coordinate system correspond to which physical spacetime points. In fact, mathematical coordinate values change through coordinate transformations, which imply automorphisms on the manifold. However, in line with *general covariance*, the fundamental principle of GRT, it is ensured that all coordinate systems should be equivalent. In other words, a spacetime point is referred to by various coordinate values, depending on the coordinate system employed. Hence, solely coordinate values lack any intrinsic or inherent physical information. They can only correspond to physical spacetime points through the metric, which provides extrinsic local information. Even Einstein himself took

[9] Not only physicists but also many philosophers concur that the metric holds a primary status as physical spatiotemporal attributes, with topology gained through the coverage of the metric [40, 50, 51]. However, there exist mathematical global properties that cannot be reduced to the metric or are anterior to it within a manifold. For an in-depth exploration of this, please refer to [49]. It is deduced that *topology is independent of locality, but it arises from metric*, thus this fact does not stand in opposition to the consensus ([49], p. 22).

2 years to fully embrace this novel spacetime perspective[10]. The fundamental essence of physical spacetime being rooted in locality would be characterized in contemporary discussions within the context of the philosophy of spacetime.

In addition, if we set STR aside and focus solely on GTR, the independence of spacetime as a distinct entity becomes uncertain. Since $R_{\mu\nu}$ and R are derived from $g_{\mu\nu}$, which describes spatiotemporal properties, it can be construed that the left-hand side of the equation delineates spacetime with its gravitational/metric field, while the right-hand side encompasses matter[11]. Admittedly, I have previously acknowledged that spacetime appears to function as a dynamic medium interacting with matter, as illustrated by Eq. (1), yet this perception might be somewhat superficial. This is because Einstein's equation allows for boundary concepts between spacetime and matter, specifically vacuum energy or dark energy mentioned earlier.

The narrative commences with the introduction of the cosmological constant. When Einstein incorporated a cosmological constant in 2, Eq. (1) underwent a slight modification:

$$R_{\mu\nu} - \frac{1}{2}g_{\mu\nu}R + \Lambda g_{\mu\nu}. \tag{3}$$

Here, Λ represents the cosmological constant, serving as a repulsive force. But how did it come about? A cosmological constant can be perceived as an additional *Lagrangian* term distinct from $\mathcal{L}_m$ for ordinary matter in the entire action, which underpins the derivation of Eqs. (1) and (3):

$$S = \int \left[\frac{1}{16\pi G}(R - 2\Lambda) + \mathcal{L}_m \right] \sqrt{-g} d\Omega. \tag{4}$$

Here, g denotes the determinant of the metric tensor $g_{\mu\nu}$ and $d\Omega = cdtdxxydz$. The second term on the right-hand side represents a mathematical degree of freedom, which can be incorporated into the *Einstein-Hilbert action* (the first term) like a constant of integration. In Eq. (1), one can set $\Lambda = 0$. As described in 2, this cosmological constant was initially intended to function as a repulsive force to counteract gravity's attractive one.

However, the crucial question pertains to the attribution of this effect within the universe—namely, is it attributed to spacetime or to matter? For instance, if one perceives a vacuum state as a system where $T_{\mu\nu} = 0$, then Λ could be attributed to an empty spacetime. Hence, $\Lambda g_{\mu\nu}$ might be viewed as an effect of vacuum energy, suggesting that spacetime possesses nongravitational energy as an inherent property. This energy facilitates the universe's expansion even in the absence of material objects or fields. However, an opposite perspective can also be taken.

---

[10] In a span of 2 years, subsequent to formulating spacetime utilizing Riemannian geometry, with the assistance of Grossman, Einstein arrived at his equation depicted as Eq. (1) [52]. Throughout these 2 years, he grappled with the hole argument—a debate on whether a point within a hole devoid of matter, traversed by a test particle, can be differentiated from another point in the same hole without metrical information. This inquiry is intricately linked to a profound query concerning whether GTR is deterministic or not [53]. The hole argument has induced various philosophical discussions up to the present day [54–56].

[11] In the realm of philosophy of spacetime, there was a discussion about whether the gravitational field should be classified as part of spacetime or as a form of matter (e.g., [40, 50, 54]).

As long as one grasps Eq. (4), Λ is not linked to the derivation of R, or the spacetime side. Hence, in Eq. (3), the cosmological constant or vacuum energy can also be attributed to the matter side, as follows:

$$R_{\mu\nu} - \frac{1}{2}g_{\mu\nu}R = 8\pi G\left(T_{\mu\nu} - \frac{\Lambda}{8\pi G}g_{\mu\nu}\right) = 8\pi G\tilde{T}_{\mu\nu}. \tag{5}$$

Mathematically, this equation can be obtained by simply moving the term $\Lambda g_{\mu\nu}$ from the left-hand side to the right-hand side. However, it can also be inferred that vacuum energy is a form of matter. In this perspective, $-\frac{\Lambda}{8\pi G}g_{\mu\nu}$ corresponds to an unusual medium with negative pressure. This negative pressure is believed to exert a repulsive force counteracting gravity throughout the entire universe, as indicated by the calculation result of Eq. (5). In essence, whichever side vacuum energy is attributed to, both interpretations hold valid physical implications.

Vacuum energy is generalized to invisible dark energy, which is present throughout the universe, spanning all spatiotemporal regions. Λ itself is not always required to be constant, and the cosmological constant faces a significant issue from a theoretical perspective in quantum physics[12]. Therefore, vacuum energy has been expanded into the concept of dark energy, which encompasses not only constant but also variable quantities to explain the universe's acceleration. However, the fundamental question remains: Should the impact of dark energy be interpreted as an attribute of spatiotemporal properties or material properties?

This *underdetermination*, or the inability to definitively judge one of the two interpretations as correct solely based on observations and data, renders the philosophy of spacetime more contentious than physics[13]. The question of whether this medium is attributed to spacetime or matter might not hold great importance, as its empirical function in explaining cosmological phenomena, such as the universe's expansion, remains consistent. However, from a philosophical standpoint, the very sameness becomes a problem. When it comes to these unknown yet physically existing entities, the demarcation between spacetime and matter is no longer distinct within the framework of GTR.

Spacetime might be inseparable not only from gravitational fields but also from matter. Einstein probably initially intended to reject spacetime without matter, or physical states with $T_{\mu\nu} = 0$, in accordance with the relationist spirit. However, he made a concession to such models as De Sitter's universe, which is one of the solutions of Eq. (3) with $T_{\mu\nu} = 0$ (but $\tilde{T}_{\mu\nu} \neq 0$) proposed by Willem de Sitter as a vacuum solution with only the cosmological constant [59–61] right after proposing the static model. Einstein suggested that behind the established geometry of spacetime, there should be some material entities [62, 63]. But Einstein presupposed matter distributed in a specific region known as "the equator," not the aforementioned invisible entities spanning all spatiotemporal regions in accordance with Mach's principles. However, this equator is not physically accessible, and

---

12 Known as the cosmological constant problem, please refer to [57, 58], for example.

13 Of course, even when restricting the discussion to physical considerations, in comparison to Eq. (1), Eq. (3) differs as a law, in that Λ influences the relationship between spatiotemporal geometry and matter, whereas Eq. (5) differs as an initial condition, in that Λ influences how the material distribution is established initially. I believe that both interpretations do not contradict the Mach principle, given that matter distributions and spatiotemporal geometry can have empirical aspects.

empirical reasons for the existence of such matter were not established[14]. Expressed from a viewpoint of philosophy of science, Einstein's presupposition was too *ad hoc* to defend Mach's principle. Consequently, the concept of empty spacetime became even more ambiguous.

The twentieth-century field theory tells us that both spacetime and matter are described by fields [35]. Quantum theory has the most specific features of fields, but even when limited to classical physics, fields took the place of matter and spacetime. In fact, as I have already shown, in GTR spacetime cannot be described without a gravitational field, and this field leads to a global structure of manifold with topology. Einstein recognized this as well:

> *If we imagine the gravitational field, i.e. the functions g_{ik} to be removed, there does not remain a space of the type (I) [Minkowski spacetime], but absolutely nothing, and also no "topological space". For the functions g_{ik} describe not only the field, but at the same time the topological and metrical structural properties of the manifold ([64], p. 155).*

Furthermore, matter can also be referred to by material fields, expressed as $T_{\mu\nu}$ or $\mathcal{L}_m$ rather than by discrete bodies. In classical physics, there are two representative fields—one is the gravitational/metric field, and the other is the electromagnetic field identified as light. Within classical physics, these fields may have been the most fundamental candidates for describing the physical world, namely the universe, rather than spacetime and matter.

Finally, Einstein understood that spacetime was a structural quality of the field, not subordinate to matter. Given the ambiguity between matter and spacetime, it may have been natural for Einstein to strive to establish a *Unified Field Theory* (UFT), which attempts to unify these fields into one, in the mid-1920s after relinquishing Mach's principle ([65]; cf. [35], p. 101)[15]. Whether or not this interpretation is accurate depends on the development of physics itself, or whether UFT will eventually be realized in the future. However, even when discussing philosophical realism, the entire field should be regarded as central entities, as expressed below:

> *There is no such thing as an empty space, i.e. a space without field. Space-time does not claim existence on its own, but only as a structural quality of the field ([64], p. 156).*

The priority of matter over spacetime or even the equivalence of both notions may be irrelevant, suggesting that only the fields exist in the world. What we perceive as spacetime and matter may merely be epistemological forms used to represent the underlying structure.

In short, when asked the question "Is spacetime real? Does spacetime exist?", the answer can be simplified to the existence of a geometric structure in which each point possesses only extrinsic local properties in relation to others, established by the gravitational field. *Exemplification* or *instantiation* of this structure is indifferent—whether

[14] For a detailed discussion on it, please refer to ([35], p. 82–85).

[15] Einstein was ultimately unable to complete this work, but the task of unifying fields, as pursued by UFT, is said to have been taken over by quantum field theory in the realm of contemporary physics.

it is attributed to pure spacetime, matter, dark energy, or any other factor[16]. What holds significance is the structure itself, referring to something present in the physical world and described mathematically [40]. This interpretation, structural realism concerning spacetime, distinguishes itself from traditional substantivalism and relationism while it gets the best of both worlds.

2.2 What is expanding?

In simple terms, an expanding universe literally means that the scale of the universe is increasing. The greater the distance between galaxies and clusters of galaxies distributed in the universe becomes, the larger the universe itself becomes. As I mentioned in 3, observations show that the redshift, which is the increase in wavelength of light reaching Earth, indicates how fast a star or galaxy from which the light was emitted is moving away. This information indirectly helps us determine the rate of expansion. Initially, I would like to accept the expansion of the universe as an empirical fact.

Well, now that the spacetime picture in GTR has been established, I will explore the theory of an expanding universe using this structural perspective. In fact, the reason behind the universe's expansion, even if it holds true, still lacks a satisfactory explanation. One model that attempts to explain the expanding universe is based on the concept of space itself. Let me elaborate on this view, which ultimately leads to the fundamental question: How did spacetime come into existence?

In order to discuss the entire universe, one needs to establish a basic model for the global universe system. In standard cosmology, the strong cosmological principle is initially applied to the system—our universe exhibits isotropy and homogeneity on a global scale[17]. Under this cosmological principle, a specific metric tensor $g_{\mu\nu}$ from Eq. (2) can be derived as the *Friedmann-Lemaître-Robertson-Walker metric* (FLRW metric)[18]:

$$ds^2 = c^2dt^2 - a^2(t)\left[\frac{dr^2}{1 - Kr^2} + r^2\left(d\theta^2 + \sin^2\theta d\phi^2\right)\right]. \qquad (6)$$

This metric is represented in a four-dimensional coordinate notation as (t, r, θ, ϕ), where the time coordinate is denoted as t, and the three-dimensional space is parameterized using polar coordinates (r, θ, ϕ). The spatial metric in three dimensions can be expressed using the constant spatial curvature K. If the universe possesses a finite and closed space, the value of K is positive, whereas if the universe has an infinite and open space, the value of K is equal to or less than zero standardly[19]. It should be noted that in the spatial coordinate system, known as the *co-moving coordinate system*, the scale factor $a(t)$ changes as proper time t for observers increases, causing expansion or

---

[16] Structural interpretations vary, and structural realism has various perspectives [66]. Structural realists are ambiguous about which physical entity is referred to by a spatiotemporal structure. However, among them Esfeld and Lam hold a clear substantivalist view that spacetime points are considered as objects, and the geometric local relations they maintain with other points themselves constitute a complete spatiotemporal structure [45].

[17] This is also one of the metaphysical assumptions of Cosmology.

[18] For example, for the derivation process, please refer to ([67], pp. 2–4).

[19] If the value of K is zero, the space becomes purely Euclidean.

contraction of the universe. That is, the scale of the spatial coordinate system changes in time for observers at rest with respect to the moving spatial coordinate system. Additionally, it is important to mention that the origin of the polar coordinates is set at the observer's position (Earth in the Milky Way galaxy).

The FLRW metric is often likened to a balloon model. It is the scale factor $a(t)$ that drives the expansion of the universe. This factor, as its name implies, serves as a kind of ruler for measuring the scale of the spatial coordinate system. When the scale factor increases, the scale of the system also increases, regardless of whether the system or the universe itself is finite or infinite. The scale of the spatial coordinate system corresponds to the scale of the universe. Thus, this coordinate system is referred to as a co-moving coordinate system. Galaxies and clusters of galaxies move in accordance with this spatial coordinate system, their positions on the system remaining unchanged, much like the marks on a balloon moving apart as the balloon expands.

This explanation of the universe's expansion is straightforward and easy to understand, but it does carry the potential for misunderstanding. The explanation, reliant on a co-moving system, seeks to offer a rationale for the behavior of the scale factor $a(t)$ by solving Eqs. (1) and (5), presupposing the isotropy and homogeneity of the universe based on the cosmological principle[20]. In the calculation, $T_{\mu\nu}$ is expressed as a combination of various ideal fluids, characterized by zero viscosity and constant densities, including dark energy, as observed from the spatial coordinate system[21]. However, if measured using a spatial unit of the co-moving coordinate system linked to the matter's scale, these densities and volumes remain constant. As I mentioned in the previous subsection, $T_{\mu\nu}$ cannot describe discrete entities like galaxies. This ideal model is insufficient to describe the actual distribution of matter.

Furthermore, this model does not provide an explanation for the fundamental reason behind the current expansion of the universe. For whether the value of $a(t)$ increases or decreases depends on particular additional assumptions, such as the initial conditions $a(0) = 0$ and $a(t) > 0$. In other words, solutions for $a(t)$ as increasing functions, signifying the universe's expansion, are obtained under the premise that the universe possesses a beginning ($t = 0$) with a singularity ($a(0) = 0$). The balloon model utilizing the FLRW metric does not inherently provide the explanation for the cause of this expansion, but just expresses the expansion itself.

The notion that outer space is expanding only presents this model on a surface level. Phenomenally in physics, the alteration in the scale of the universe is merely assumed. This model employs a co-moving coordinate system to depict this alteration and describes the expansion as if the spatial unit were increasing due to the growth of $a(t)$. It is essential to bear in mind that this spatial depiction is confined to the co-moving coordinate system. The positions of galaxies in the universe are not changed by expanding spatial forces or similar factors. Observers arbitrarily fix the spatial coordinate system on their positions. The expansion of the universe is not a consequence of the space itself expanding, but rather it is caused by other factors such as energy, entropy, or other elements that are not accounted for in this ideal symmetric model.

---

[20] In general, concrete solutions for $a(t)$ involving functions of t can be theoretically derived by solving Friedmann equations, which are obtained by applying Einstein's equation to the FLRW metric.

[21] Even if the universe's scale were measured using a constant unit of length, the densities of these fluids decrease as the universe expands due to the increase in the volume of matter.

Expanding space is merely a metaphor. The idea that the spatial coordinate system expands in accordance with the increase of the time parameter is a mathematical representation rather than a physical phenomenon. In this context, space does not possess the nature of a substance like absolute spacetime. While General Relativity Theory (GRT) describes the dynamic nature of spacetime, it does not attribute material characteristics to it, as illustrated by the traffic light example mentioned in the previous subsection.

Let us revisit the characteristics of the absolute spacetime that Newton presupposed:

> If absolute spacetime is considered a substance, these points labeled (x, y, z) *persist through time t, remaining the same across different moments in time ([41], p. 56).*

If there is a concept of expanding space in the universe, each spatial point labeled (r, θ, ϕ), which remains constant over time t, appears to move away from one another, as if each spatial coordinate value corresponds to an individual substance. However, this depiction is primarily a lengthy assumption.

For instance, in GTR, space points are not the same across different times. In the co-moving coordinate system, let us consider a space point labeled p with spatial coordinates $\left(x_p, y_p, z_p\right)$ at time coordinate value t_1 in the FLRW metric. This space point is not identical to another point labeled p at the same coordinate system but at a different time coordinate value t_2, as space points do not persist through time. Point p at t_1 and point p at t_2 coincidentally possess the same spatial coordinates within the framework of the co-moving coordinate system (**Figure 1**). According to the principle of general covariance, even if a different coordinate system were applied to these two points, physical phenomena should be described in the same manner. Hence, there is no need to assert their identity since these two spacetime points labeled $\left(t_1, x_p, y_p, z_p\right)$ and $\left(t_2, x_p, y_p, z_p\right)$ in the co-moving system are assigned distinct spatial coordinates in the new coordinate system.

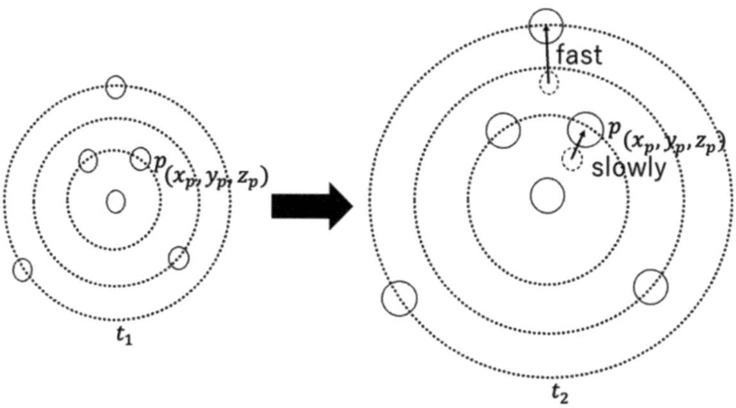

Figure 1.
The explanation of an expanding universe with a co-moving coordinate system.

Upon investigation, the co-moving coordinate system (t, r, θ, ϕ) can be converted into the static coordinate system (t, r', θ, ϕ), keeping the scale of the spatial coordinate system unaltered through the coordinate transformation $(r' = a(t)r)$:

$$ds^2 = \left(c^2 - H^2(t)r'^2\right)dt^2 + 2H(t)dtdr' - \frac{dr'^2}{1 - \frac{Kr'^2}{a^2(t)}} - r'^2\left(d\theta^2 + \sin^2\theta d\phi^2\right). \quad (7)$$

Here, $H(t)$ represents the Hubble constant, and it can be expressed using $a(t)$ as $H(t) = \frac{\dot{a}(t)}{a(t)}$ (where $\dot{a}(t)$ is the first derivative of $a(t)$ with respect to t).

This coordinate system, with the term *dtdr* present, is not synchronized with respect to time unless $H(t) = 0$. Consequently, the progression of time varies at each point in space, and defining simultaneity becomes impossible.

This new static coordinate system (t, r', θ, ϕ) seems quite complex, as time t from the co-moving coordinate system is still in use, but it must provide a framework to describe physical phenomena. General covariance allows this coordinate system to be equivalent to the original one. It is certain that in the static coordinate system, a spatial component (r', θ, ϕ) does not expand. As time t increases, the nonzero spatial curvature $\frac{K}{a^2(t)}$ can change[22], but the scale of the spatial coordinate system remains constant. In this coordinate system, space never expands; matter only moves away. A space point labeled p at time t_1 in the co-moving system and another at time t_2 in the same system are labeled different spatial coordinate values in the static coordinate system, respectively.

However, the co-moving coordinate system might be more than just a convenient tool for describing the universe's system. Each spatial coordinate value in the co-moving coordinate system corresponds to a physical position. It directly corresponds to the position of each galaxy. If the universe is isotropic and homogeneous, using this coordinate system is very natural and logical. In this sense, the co-moving coordinate system might be the most suitable model for representing spacetime itself.

The co-moving coordinate system saves the cosmological principle. Hubble also demonstrated through observational data on redshift that the farther galaxies are from Earth, the faster they move away, with velocities proportional to their distance from Earth [24]. Intuitively, these results seem to suggest that all celestial bodies are moving away from the center of the universe, namely our Milky Way galaxy, considered seemingly a special place, which contradicts the cosmological principle. However, if the universe truly expands like a balloon, these data are consistent with the cosmological principle as **Figure 1** shows it. Thus, if there is no expanding space behind matter, there is no explanation for why the universe is expanding in this manner. This *reduction to the absurd* resembles a *no-miracle argument*, asserting that expanding space is real which appeals to *Inference to the Best Explanation* [68], even though the distinction between material positions and space points, or spacetime and matter, remains ambiguous. The balloon model not only explains the way phenomena occur but also provides a clear worldview on the expanding space concept.

However, the structural interpretations discussed in the previous subsection refute the notion of expanding space. Certainly, the spatial coordinate values in the co-moving coordinate system might refer to the physical positions of galaxies. The co-moving coordinate system could suggest the existence of expanding space behind matter. But let

---

22 space approaches flatness: $\frac{K}{a^2(t)} = 0$.

us remember the perspective of structural realism. Structural realism does not concern itself with what the concrete attribution of the spatiotemporal structure is.

In this context, it is crucial to recall that each spacetime point lacks intrinsic physical properties on its own. They possess only extrinsic or relational local properties provided by the metric. A coordinate value devoid of metric information cannot uniquely refer to a specific spacetime point, even though it refers to a position within the universe.

For a structural realist, both the co-moving coordinate system and the static coordinate system are merely different forms to describe the same spatiotemporal structure. The notion that two spatial points are uniquely identified or persist through time is not upheld by structural realism, which is the reasonable interpretation within classical spacetime, particularly within GTR. The co-moving coordinate system is essentially a convenient mathematical aid, but the entirety of four-dimensional spacetime can be described using other complex coordinate systems, such as the static coordinate system.

Structural realism concerning spacetime informs us that the abstract mathematical representation, like the FLRW metric derived from gravitational fields, undoubtedly refers to something physical in the universe, but it does not assume a specific form with size and shape, akin to a physical substance like the expanding universe. Many mathematical tools might represent this structure, but attributing a single physical entity exclusively to any one of them is not in line with the essence of structural realism, even though the distinction between spacetime and matter might be ambiguously claimed by structural realists. Expanding space is not endorsed by the structural worldview [22, 69].

If space were to expand as a substance, this expansion should have had a starting point. Assuming a singularity like $a(0) = 0$ when $t = 0$ signifies that the universe, encompassing all material entities, came into being from a size of zero at $t = 0$, in accordance with classical perspectives[23]. Thus, the beginning of the universe aligns with the beginning of spacetime. In this interpretation, the void from which the universe occurred is not equivalent to space itself, but rather it denotes nothingness or something distinct from space devoid of matter.

However, the structural interpretation that negates the existence of an expanding universe furnishes an alternate picture of the universe's beginning:

> Even if the universe contracted to a point of zero, space did not similarly contract. Conse-quently, *at the beginning, a spatiotemporal structure could be postulated to exist around the singularity [69].*

In this picture, the vacuum underlying the universe, which indicates the region distributed by matter, might be synonymous with spacetime devoid of matter. This is because spatiotemporal structures can exist independently of matter, akin to Minkowski metric and Schwarzschild metric, which serve as vacuum solutions of Eq. (1). In this sense, spacetime constitutes the universal background of the universe from the perspective of structural realism.

---

[23] Strictly speaking, this depiction might be relevant to a finite universe with a positive K. However, at present, I consider an infinitely scaled spacetime with $K \leqq 0$, which can converge to zero, based on the belief that the size of material parts is finite.

This is precisely the answer I aim to present to the fundamental question. Structural realism concerning spacetime does not accept the notion of the beginning of spacetime in cosmology. It is quite intriguing that this audacious assertion is derived not from QGT addressing the question, but from philosophical concepts rooted in GTR.

3. Are we going to the non-spacetime picture?

Spacetime is often said to emerge from more fundamental entities in Quantum Gravity Theory (QGT). In simpler terms, spatial properties, such as distance and continuity, cease to hold true in microscale regions like the Planck scale. This arises because the manner in which gravitational fields integrate quantum effects within these domains remains unclear. Consequently, the notion that spacetime is not fundamental necessitates a reconsideration of conventional empirical concepts such as observations and causality. These concepts must be interpreted in a manner that accounts for the absence of spacetime.

This task also influences both physics and philosophy. The novel worldview has the potential to induce the next scientific revolution, as physics is on the verge of revisiting the fundamental nature of spacetime since the advent of general relativity [31]. This, in essence, is the primary role of physics, not metaphysics nor philosophy. But this challenge also prompts philosophers to contemplate the existence of *empirical coherence* [70] and the distinction between mathematical entities and physical entities, or *physical salience* [71] in QGT.

If Quantum Gravity Theory (QGT) can formulate consistent theories that integrate quantum theory and GTR, or formalize GTR in microscale regions, this impending revolution could present a coherent framework elucidating the emergence of spacetime. This framework would address the question of which aspects of a spatiotemporal structure are taken to these microscale regions.

The term "emergence" is highly ambiguous and controversial. In physics, emergence pertains not only to entities and properties but also to phenomena and theories that are "novel" and "robust" in relation to more fundamental ones [72]. In other words, spacetime is more than just a composition of constituent elements; it encompasses spatiotemporal properties that cannot be explained solely by the functions of these constituents. Moreover, the discourse concerning the emergence of spacetime takes on different aspects within various types of QGTs.

In the previous section, the conclusion was reached that the structural realism of spacetime, as derived from GTR, contradicts the notion of expanding space within relativistic cosmology. This perspective strongly asserts that space is distinct from a substance-like entity which came into being from the singularity. It does not endorse the notion that a spatiotemporal structure could not disappear at the universe's birth. If this interpretation is valid, what lessons can we draw from Quantum cosmology's exploration of the early micro universe?

QGT is closely related to Quantum cosmology because, following the standard Big Bang cosmology, the early universe contracted into a small size, irrespective of whether space also underwent contraction. This implies that some of the QGTs describe the early universe near the singularity, searching for the origin of the universe or spacetime. Toward the conclusion of this chapter, I will illustrate how the structural view addresses this non-spacetime theory in Quantum cosmology.

However, it is worth noting that structural realism about spacetime is rooted in GTR, whereas QGT does not necessarily presuppose GTR in macroscopic regions. Indeed, *Super String Theory* (SST), one of the QGTs, operates with a distinct classical field theory that becomes approximately consistent with GRT under high-energy conditions. This theory establishes a definite spatiotemporal structure as a background for a string or other material entities by disassociating gravity from spacetime, quantifying gravitational fields along with other fields like electromagnetic ones. SST might be the most coherent QGT, yet its concept of spacetime starkly diverges from the structural interpretation that equates spatiotemporal structures with gravitational fields, leaving aside the fact that, in SST, spacetime extends to 11 dimensions rather than the usual 4.

Hence, I find it exceedingly challenging to establish a connection between the notion of "spacetime" in SST and the spacetime discussed in the previous section. However, SST can be viewed as a theory addressing the emergence of spacetime, which is recognized as *T-duality*. In contrast to ordinary spatial domains, whose sizes are quantifiable, the *target space* employed as a backdrop in SST operates on a lower level of observability and can possess physically equivalent yet distinct length scales, a perplexing concept [73]. In the microscale regions of this target space, concepts like distance lose their clear definition. This serves as an illustration of more fundamental entities exhibiting non-spatiotemporal attributes.

As other more fundamental entities, which spacetime is emergent from, concepts, such as the *spin network* or the *spin foam* in *Loop Quantum Gravity* (LQG) and the *causal set* in *Causal Set Theory* (CST), have arisen. LQG and CST share the common goal of elucidating the constituents of spacetime itself—akin to the notion of a space or spacetime atom. The spin network characterizes nodes connected with edges within an abstract graph, representing discrete volume and surface elements of the space atom. The causal set, on the other hand, depicts spacetime points arranged within a causal relation that conveys a temporal dimension in a coherent manner, as if all points collectively compose a unified space. These depictions seemingly suggest that spacetime possesses substance-like attributes, reminiscent of the old substantivalism, which appears to be at odds with the structural perspective. It is worth noting that both LQG and CST reflect theoretical structures within the framework of GTR[24].

Especially, LQG directly quantizes metric, but there is no isomorphic correspondence between structure of spin network and the spatiotemporal one because through quantization processes, volume or surface area, from which continuity of space is emergent, is defined and spatiotemporal properties such as locality are lost [75][25]. Hence, there is a mathematical discontinuity between spin network and spacetime, which is the cause of emergence.

However, a spatiotemporal structure does not change into spin network through quantization. Indeed, LQG begins with the Einstein-Hilbert action calculated from $g_{\mu\nu}$ in Eq. (4), a spatiotemporal structure. To enable canonical quantization, the action needs to be transformed into the Palatini-Holst action [77], which is classically equivalent to it. When defining the configuration variable A_a^i called the *Ashtekar variables*, its *canonical conjugate momentum* is expressed using the scalar density of the *triad field* $\tilde{E}_a^i$. With this Astekar representation, the new action is given as below:

[24] As to CST, please see [74].

[25] As to the emergence of locality, please see [70, 76].

$$S = \frac{1}{8\pi G\beta}\int d^4x\left\{\tilde{E}_i^a\left(\mathcal{L}_t A_a^i\right) + N\epsilon_{ijk}\tilde{E}_i^a\tilde{E}_j^b F_{ab}^k + N^a\tilde{E}_i^b F_{ab}^i + \lambda^i\left(D_a\tilde{E}_i^a\right)\right\}. \tag{8}$$

Here, β is the *Barbero-Immirzi* parameter, N, N^a, and λ^j are undetermined constants of the Lagrangian, $\mathcal{L}_t$ represents the *Lie derivative*, ε_{ijk} is the *Levi-Civita epsilon* (antisymmetric tensor according to Einstein's summation convention), F_{ab}^i is the differential form of A_a^i given analogously by *Young-Mills theory* for the electromagnetic field, and D_a represents the covariant derivative[26]. This is simply a change of variables, and Eq. (8) still describes a classical structure. Furthermore, even when canonical variables $\left(A_a^i, \tilde{E}_i^a\right)$ are quantized, there exists an isomorphic correspondence between structures, quantum numbers, and classical numbers. Quantized spacetime is mathematically connected to classical spacetime through the structural interpretation.

The structure of a spin network can be derived from the calculation of quantized spacetime/gravitational field, not from classical spacetime. Through the quantization of Eq. (8), a wave function Φ with a variable of A_a^i is associated in relation to other known quantizations of different fields. However, to establish conditions without spacetime geometry as a background, in which physical quantities can be defined, a more abstract and complex state space is required. It is the spin network which is described in this unique space, and the discrete physical quantities are variables related to quantized original spacetime, namely the volume and surface areas of units of space.

Therefore, the spin network could be a space/spacetime atom, but not space/spacetime itself. It is because space atoms are described in another space detached from a metrical system of Eq. (8) in which spacetime itself is described. This abstract system does not constitute a spatiotemporal region. Two different foundations have different concepts. Consequently, spatiotemporal properties emerge independently of what a spin network describes, to some extent [75].

Ultimately, could QGT potentially describe the origin of spacetime itself in the future? If we simplify the idea, if an expanding space contracts into a singularity along with matter[27], QGT might describe a space atom of extremely small size. However, this concept of the size of a space atom is developed as a spin network, within a different abstract system. As long as there remains a conceptual discreteness between the spin network and spacetime, even space atoms resembling a substance could remain consistent with the structural realism about spacetime. This is because the latter is independent of the properties of the former.

Is this abstract foundational space a physical region? In physics, there exist various types of space beyond the four-dimensional spacetime, such as high-dimensional space, topological space, and phase space. Let us call these abstract space non-spacetime now. The concept of state/Hilbert space was a conceptual or mathematical one used to describe the behavior of a system. However, QGT blurs the boundaries of this abstract role even further. If spin networks and causal sets are indeed real as space atoms, where would they exist? Does this question make sense without a spatiotemporal structure? Can we entertain the idea of a universe or physical regions existing out of spacetime?

---

[26] A sequence of detailed process and each variable depends largely on ([78], pp. 91–93).

[27] A model was proposed that the compressed universe transitions from contraction to expansion due to the quantum effect of repulsive gravity to avoid a singularity, inducing Big Bounce in QGT [79].

If we consider physical regions to encompass the entire universe, does the concept of emergent spacetime allow for the inclusion of non-space regions as part of the physical ones, rather than mere mathematical models or tools? This notion has already been introduced in modern physics: for example, when quantizing the electromagnetic field, photons oscillate not in physical space but within an abstract space formed by the physical quantities of the field itself [80]. However, virtual particles like photons correspond to fields that exist within spacetime, and the empirical aspects or phenomena associated with these particles occur within a spacetime region.

When it comes to space atoms, the possibility of non-spacetime regions comes into play. Causality might serve as a bridge connecting the spatiotemporal universe to a non-spatiotemporal one, potentially through the context of the big bang. To consider the behaviors of spin networks and causal sets as genuine phenomena, it might be necessary to expand the scope of physics even further!

Acknowledgements

This work was supported by Research Grants of Suntory Foundation in Japan. Thank you very much.

Author details

Sho Fujita
Department of Complex Systems Science, Nagoya University Graduate School of Informatics, Nagoya-City, Aichi, Japan

*Address all correspondence to: fujita.sho.p6@a.mail.nagoya-u.ac.jp

References

[1] Scully S. Hesiod's Theogony, from Babylonian Creation Myths to Paradise Lost. New York: Oxford University Press; 2015

[2] Evans J. The History and Practice of Ancient Astronomy. New York: Oxford University Press; 1998

[3] Neugebauer OE. A History of Ancient Mathematical Astronomy. Berlin: Springer-Verlag; 1975

[4] Aristotle. De Caelo. (350 BC), Leggatt S. (trans and ed) on the Heavens I and II, Warminster: Aris & Phillips Ltd; 1995

[5] Copernicus, Nicolaus. "Nicolai Copernici Torinensis De revolutionibus orbium coelestium, Libri VI" (1543), Wallis CG (trans) On the Revolutions of Heavenly Spheres (Great Minds Series). New York: Prometheus; 1995

[6] Brahe T. Tychonis Brahe Astronomiæinstauratæmechanica. (1602). Marrickville: Wentworth Press; 2016

[7] Kepler J. Astronomia nova, (1609). In: Do-nahue WH. (trans). New Astronomy, Cambridge: Cambridge University Press; 1992

[8] Drake S, Mahoney MS. Telescopes, Tides, and Tactics: A Galilean Dialogue about the Starry Messenger and Systems of the World. London: University Of Chicago Press; 1983

[9] Galilei G. Sidereus Nuncius, (1610). In: Helden AV. (tans and ed). Sidereus Nuncius, or the Sidereal Messenger, 2nd edition; Chicago: University of Chicago Press; 2016

[10] Galilei G. Dialogo sopra i due massimi sistemi del mondo, (1632), In: Drake S. (trans & ed). Dialogue Concerning the Two Chief World Systems, Ptolemaic and Copernican. 2nd Revised edition. California: University of California Press; 1962

[11] Galilei G. Discorsi e dimostrazioni-matematiche intorno a due nuove scien-ze, (1638). In: Crew, H, de Salvio A. (trans). Dialogues Concerning Two New Sciences. Glasgow, England: Good Press; 2021

[12] Descartes, René. "Discours de la Méthode Pour bien conduire sa raison, et chercher la vérité dans les sciences" (1637), Maclean I. (trans) A Discourse on the Method of Correctly Conducting One's Reason and Seeking Truth in the Sciences. New York: Oxford University Press; 2006

[13] Newton I. Philosophiæ Naturalis Principia Mathematica, (1687). In: Cohen IB, Whitman A (trans). Contributed by Budenz J The Principia: Mathematical Principles of Natural Philosophy, California: University of California Press; 1999

[14] Clarke S, Leibniz GW, Newton I. The Leibniz-Clarke Correspondence: Together with Extracts from Newton's Principia and Opticks. In: Alexander HG, editor. New York: Barnes & Noble; 1956

[15] Leibniz GW. La Monadologie, (1714). In: Latta R, (trans). The Monadology. California: Create Space Independent Publishing Platform; 2015

[16] Leibniz GW. Discourse on metaphysics (1686). In: Loemker LE, editor. Philosophi-cal Papers and Letters: A Selection. Dor-drecht: Springer; 1976. pp. 303-330

[17] Leibniz GW. Nouveaux essais sur l'en-tendement humain, (1765). In:

Remnant P, Bennett J (trans and eds). New Essays on Human Understanding. Cambridge: Cambridge University Press; .1981

[18] Leibniz, GW, Clarke S. Correspondence. In: Ariew R. editor. Indianapolis: Hackett Publishing Company; 2000

[19] Kant I. Kritik der reinen Vernunft. (1787). In: Paul Guyer Allen W. Wood (trans and ed). Critique of Pure Reason. Cam-bridge: Cambridge University Press; 1998

[20] Mach E. Mechanik in ihrer Entwicklung; historisch-kritisch dargestellt, (1883). In: McCormack TJ, (trans). The Science of Mechanics: A Critical and Historical Account of its Development (1893). Chicago: The Open Court Publishing Company; 6 edition; 1960

[21] Einstein A. Ernst Mach. Physikalische Zeitschrift. 1916;**17**: 101-104

[22] Fujita S. An interpretation on structural realism to space-time used in big-bang cosmology: Is space really expanding? Journal of the Japan Association for Philosophy of Science "Kagaku Kisoron Kenkyu" in Japanese. 2017;**44**(1–2):1-14

[23] Einstein A. Kosmologische Betrachtungen zur allgemeinen Relativitatstheories. Sitzungsberichte der Königlich Preuβischen Akademie der Wissenschaften. 1917:142-152

[24] Hubble E. A relation between distance and radial velocity among extragalactic nebulae. National Academy of Sciences. 1929;**15**(3):168-173

[25] Gamov G. The origin of elements and the separation of galaxies. Physics Review. 1948;**74**(4):505-506

[26] Penzias A, Wilson R. A measurement of excess antenna temperature at 4080 mc/s. Astrophysical Journal. 1965;**142**:419-421

[27] Guth AH. The inflationary universe: A possible solution to the horizon and flatness problems. Physical Review D. 1981;**23**:347-356

[28] Riess AG, Filippenko AV, Challis P, et al. Observational evidence from supernovae for an accelerating universe and a cosmological constant. The Astronomical Journal. 1998;**116**(3): 1009-1038

[29] Bondi H, Gold T. The steady-state theory of the expanding universe. Monthly Notices of the Royal Astronomical Society. 1948;**108**:252-270

[30] Hoyle F. A new model for the expanding universe. Monthly Notices of the Royal Astronomical Society. 1948; **108**:372-382

[31] Kuhn TS. The Structure of Scientific Revolutions. Chicago: University of Chicago; 1962

[32] Einstein A. Zur Elektrodynamik bewegter Körper. Annalen der Physik. 1905;**322**(10):891-921

[33] Einstein A. Lichtgeschwindigkeit und Statik des Gravitations-feldes. Annalen der Physik. 1912;**343**(7):355-369

[34] Einstein A. Letter to Schwarzschild. In: Cao TY, editor. Conceptual Devel-opments of 20th Century Field Theories. Cambridge: Cambridge University Press; 1997, p. 67

[35] Cao TY. Conceptual Developments of 20th Century Field Theories. Cambridge: Cambridge University Press; 1997

[36] Einstein A. Die Grundlage der allgemeinen Relativitätstheorie. Annalen der Physik. 1916;**354**(7):769-822

[37] Einstein A, Grossmann M. Entwurf einer verallgemeinerten Relativitätstheorie und einer Theorie der Gravitation. Zeitschrift fuer Mathematik und Physik. 1913;**62**:225-261

[38] Einstein A. Ether and the theory of rela-tivity. In Janssen M, Norton JD, Renn J, Sauer T, Stachel J, editors. The Genesis of General Relativity (Boston Studies in the Philosophy of Science 250). Vol 3. Dordrecht: Springer; 2007. pp. 613-619

[39] Maudlin T. The essence of space time. In: Proceedings of the Biennial Meeting of the Philosophy of Science Association. Vol. 2. Cambridge: Cambridge University Press; 1988. pp. 82-91

[40] Dorato M. Substantivalism, relationism, and structural spacetime realism. Foundations of Physics. 2000; **30**:1605-1628

[41] Maudlin T. Philosophy of Physics: Space and Time. Princeton & Oxford: Princeton University Press; 2012

[42] Newton I. Unpublished Scientific Papers of Isaac Newton. Rupert Hall and Marie Boas Hall (trans). Cambridge: Cambridge University Press. (1962).

[43] Descartes R. Principia philosophiae. (1644), Reynolds B. (trans). Principles of Philosophy. Lewiston, NY: E. Mellen Press; 1988

[44] Dorato M. Is structural spacetime realism relationism in disguise? The supererogatory nature of the substantivalism/relationism debate. In: Dick D, editor. The Ontology of Spacetime II (Philosophy and Foundations of Physics Vol. 4). Amsterdam: Elsevier; 2008. pp.17-37

[45] Esfeld M, Lam V. Moderate structural realism about space-time. Synthese. 2008;**160**(1):27-46

[46] Lam V, Esfeld M. The structural metaphysics of quantum theory and general relativity. Journal for General Philosophy of Science. 2012;**43**(2): 243-258

[47] Pauri M. Oggettivitàe realtà. In: Minazzi F (ed) L'oggettivitàdella conoscenza scientifica. Milano: Angeli; 1996 pp. 100-112

[48] Wüthrich C. Challenging the spacetime structuralist. Philosophy of Science. 2009;**76**(5):1039-1051

[49] Fujita S. What is essential to constitute a physical realm? Hints from a spatiotem-poral structure. In: Stefanov AS, Donald GD-M, editors, Spacetime Conference 2022. Montréal: Minkowski Institute Press; 2023. pp. 3-31

[50] Hoefer C. The metaphysics of space-time substantivalism. The Journal of Philosophy. 1996;**93**(1):5-27

[51] Maudlin T. Substances and space-time: What Aristotle would have said to Einstein. Studies in History and Philosophy of Science. 1990;**21**:531-561

[52] Einstein A. Die Feldgleichungen der Gravitation. Sitzungsberichte der Königlich Preuβischen Akademie der Wissenschaften (Berlin). 1915:844-847

[53] Einstein A. Die formale Grundlage der allgemeinen Relativitätstheorie. Sitzungsberichte der Königlich Preuβischen Akademie der Wissenschaften (Berlin). 1914: 1030-1085

[54] Earman J, Norton J. What price substantivalism: The hole story. British Journal for the Philosophy of Science. 1987;**38**:515-525

[55] Stachel J. The relations between things versus the things between relations: The deeper meaning of the hole argument. In: Malament DB, editor. Reading Natural Philosophy: Essays in the History and Philosophy of Science and Mathematics. Chicago and La Salle, Illinois: Open Court; 2002. pp. 231-266

[56] Stachel J. The hole argument and some physical and philosophical implications. Living Reviews in Relativity. 2014;**17**:1

[57] Peebles PJE, Ratra B. The cosmological constant and dark energy. Reviews of Modern Physics. 2003;**75**(2): 559-606

[58] Weinberg S. The cosmological constant problem. Reviews of Modern Physics. 1989;**61**(1):1-23

[59] de Sitter W. On Einstein's Theory of Gravitation and its Astronomical Consequences. Third Paper, Monthly Notices of the Royal Astronomical Society. 1917;**78**(1):3-28.

[60] de Sitter W. Further remarks on the solutions of the field-equations of the Einstein's theory of gravitation. In: Huygens Institute-Royal Netherlands Academy of Arts and Sciences (KNAW), Proceedings, 20 II. Amsterdam; 1918. pp. 1309-1312

[61] de Sitter W. On the curvature of space. In: Huygens Institute-Royal Netherlands Academy of Arts and Sciences (KNAW), Proceedings, 20 I. Amsterdam; 1918. pp. 229-243

[62] Einstein A. Kritisches zu einer von Hrn. de Sitter gegebenen Lösung der Gravitationsgleichungen, Sitzungsberichte der Königlich Preuβische Akademie der Wissenschaften (Berlin). 1918:270-272

[63] Einstein A. The Einstein-De Sitter-Weyl-Klein Debate. In: Schulmann R, Kox AJ, Janssen M, Illy J, editors, The Collected Pa-pers of Albert Einstein. Vol. 8, The Berlin Years: Correspondence, 1914-1918. Princeton, NJ:: Princeton University Press; 1998. pp. 351-357

[64] Einstein A. Relativity and the problem of space. In: Relativity: The Special and the General Theory. New York: Crown; 1961

[65] Einstein A. Unified Field Theory of Grav-itation and Electricity. In: Unzicker A, Case T. (trans and eds) Session Report of the Prussian Academy of Sciences, Translation of Einstein's At-tempt of a Unified Field Theory with Teleparallelism 2005:2-6, arXiv:physics/ 0503046 [physics.hist-ph]. 1925; pp. 414-419

[66] Landry EM, Rickles DP. editors. Structural Realism: Structure, Object, and Causality (the Western Ontario Series in Philosophy of Science Book 77). Dordrecht: Springer; 2012

[67] Weinberg S. Cosmology. Oxford: Oxford University Press; 1984

[68] Putnam HW. Mathematics, Matter and Method. Cambridge: Cambridge University Press; 1975

[69] Fujita S. The Relationship Realism of Spacetime and Contemporary Physics [doctoral thesis] (translated from Japanese). Japan: Osaka University Graduate School of Human Sciences; 2019. DOI: 10.18910/72462

[70] Huggett N, Wüthrich C. Emergent spacetime and empirical (in)coherence.

Studies in the History and Philosophy of Modern Physics. 2013;**44**(3):276-285

[71] Maudlin T. Completeness, supervenience, and ontology. Journal of Physics A: Mathematical and Theoretical. 2007;**40**(12):3151-3171

[72] Butterfield J. Emergence, reduction and supervenience: A varied landscape. Foundations of Physics. 2011;**41**(6): 920-959

[73] Huggett N. Target space ≠ space. Studies in History and Philosophy of Science Part B: Studies in History and Philosophy of Modern Physics. 2017;**59**: 81-88

[74] Wüthrich C. The structure of causal sets. Journal for General Philosophy of Science. 2012;**43**(2):223-241

[75] Fujita S. Emergence of spacetime from structural interpretations: Why is spacetime's atom not spacetime itself? Journal of the Japan Association for Philosophy of Science "Kagaku Kisoron Kenkyu" in Japanese. 2020;**48**(1):1-19

[76] Wüthrich C. The emergence of space and time. In: Gibb S, Hendry RF, Lancaster T, editors. Routledge Handbook of Emergence. 1st ed. London: Routledge; 2019

[77] Holst S. Barbero's Hamiltonian derived from a generalaized Hilbert-Palatini action. Physical Review D: Part Fields. 1996;**53**:5966-5969. DOI: 10.1103/physrevd.53.5966

[78] Gambini R, Pullin J. A First Course in Loop Quantum Gravity. Oxford: Oxford University Press; 2011

[79] Bojowald M. Follow the bouncing universe. Scientific American. 2008;**299**(4):44-51

[80] Fujita S. Abstractness for nominalists: Physical worlds outside of the spacetime realm (translated from Japanese). Contemporary and Applied Philosophy. 2021;**13**:141-173. DOI: 10.14989/266868

Section 4

Some Cosmological Scenarios

Chapter 4

Cosmology with Time-Varying Curvature: A Summary

John C. Botke

Abstract

We present a new model of cosmology based on the idea that the curvature of the universe varies with time. This model predicts a present-day exponential acceleration of the expansion of the universe without reference to a cosmological constant. We propose a new model of nucleosynthesis that accounts for the existence of cosmic structures, provides a solution to the matter/antimatter asymmetry problem, and explains the origin of the CMB. We show that this process was regulated by a vacuum imprint that came into existence during an initial Planck era inflation. One consequence is that all cosmic structures came into their final form with their present-day mass more or less simultaneously at a time of about 3×10^{16} s. This explains the recent findings by the James Webb telescope of very large galaxies dating from that epoch. We show that the stability of galaxies and galaxy clusters demands that all galaxies must have developed supermassive black holes at that time. We go on to show that the phenomena attributed to dark matter are actually consequences of vacuum energy, that the conventional model for the CMB spectrum is wrong and finally, we resolve the Hubble tension problem.

Keywords: evolution of the universe, inflation, big bang nucleosynthesis, cosmic microwave background, origin of cosmic structures, dark matter, dark energy, supermassive black holes, baryonic acoustic oscillations, Hubble tension

1. Introduction

It has long been thought that the ΛCDM model was the final answer to the problem of the evolution of the universe and that only a few tweaks would be needed to complete our understanding of cosmology. In recent years, however, researchers have begun to realize that the standard model has serious problems that cannot be resolved by making minor adjustments. A number of papers (see e.g. [1, 2]) have recently been published that suggest that the standard model might need to be scraped in favor of new physics. In these notes, we summarize just such a model. This new model that makes a major departure from the standard model in terms of its viewpoint of the evolution of the universe while making detailed predictions that solve a large number of the outstanding problems of cosmology.

We will begin with a brief review the assumptions underlying the standard model and contrast these with those of the new model. First, the standard model assumes that the universe is homogeneous and isotropic on large scales. In the new model, we also make that assumption with, however, a major difference. Whereas the standard model asserts that the curvature of spacetime is constant, we assert that it varies with time and the consequences of that are huge. With time-varying curvature, the universe must undergo a present-day exponential acceleration of its expansion that has nothing to do with a cosmological constant. According to [1], one of the major problems of the ΛCDM model is that it cannot explain the magnitude of the cosmological constant. That of course, is not a problem in the new model because the cosmological constant simply does not exist.

Another fundamental problem with ΛCDM is that it cannot explain the "coincidence" that the present-day energy densities of the vacuum and matter are roughly the same whereas at the time of nucleosynthesis, they differed by many orders of magnitude. The new model demands that such a disparity must exist and, in fact, makes an accurate prediction of its magnitude.

The second major assumption of the standard model is that the matter distribution of the early universe was uniform. The most striking feature of the universe we observe is its fantastic degree of organization on all scales from stars up to the cosmic web. The standard model asserts that early quantum fluctuations seeded later accretions that became the structures but it is clear that this organization could not have been the result of random events so some organizing principle must have been responsible. Further, we will see that the scale of the largest structures demands that the definition of these structures came about in the absence of causality. The new model solves this problem by asserting that a blueprint that came into existence during an initial Planck era inflation later regulated nucleosynthesis in such a manner as to create all the cosmic structures we now observe.

A third major assumption of the standard model is that all cosmic structures are a consequence of accretion. We will show instead that when vacuum energy is included in Einstein's equations, accretive structure formation initiated by small perturbations is impossible. In the new model, it was the imprint expressing itself by regulating the creation of matter that was responsible for the structures. We go on to show that after nucleosynthesis, under the influence of initially the expansion and later gravitation, the initial matter distributions became fully formed stars, galaxies, galaxy clusters, and superclusters more or less simultaneously at around 3×10^{16}s or $z \geq 6$. This is an important point because it explains the recent findings of the James Webb telescope of large fully formed galaxies with ages too great to be explained by accretion. We also show that as part of this process, all galaxies must have formed supermassive black holes because they would have otherwise undergone free-fall collapse and that radiation from galaxies interior to galaxy clusters was the agent that prevented the clusters from also undergoing free-fall collapse.

Another major problem with the standard model is that it cannot explain the matter/antimatter asymmetry of the universe which the new model does.

The chapter is a distillation of the results presented in the series of papers listed in the reference section. We ask the reader to refer to those papers for detailed discussions and references to methods and data.

One final point is that this model should be considered as a whole instead of the collection of disjoint pieces which is the case with the standard model. The present-day Hubble constant, the present-day CMB temperature, and one other parameter associated with neutron decay, fix all the parameters of the model. There are no other parameters and, unlike the situation with the standard model, only one evolution is possible.

2. Origins and the Planck era

We will begin with an observation concerning the CMB spectrum that will lead us back to the origins of everything. In spite of the claims of the standard model, superclusters are the only structures in existence large enough to be responsible for the large peaks of the CMB spectrum. In our original paper [3], we argued that superclusters were actually responsible but we have since realized [4] that the process that created the superclusters simultaneously created the CMB spectrum with peaks of the same physical dimensions. We will have more to say about this later but the point here is that superclusters in some form must have been in existence at the time of recombination. We also know that the CMB spectrum peaks were vastly too large to be explained by any causal process. Starting at the time of recombination and moving backward, since nothing much happened between the time of recombination and nucleosynthesis, it is natural to suppose that superclusters and the CMB spectrum originated during nucleosynthesis and, indeed, their material manifestation did originate at that time. Because of causality, however, the blueprint that regulated their creation must have come into existence even earlier. The only possible conclusion is that, even though causality is a cornerstone of physics that applies to most of the history of the universe, there must have been an epoch at the beginning during which it had no meaning. The final jump backward is to a Planck-era inflation.

Imagine that we are interested in measuring the duration of some event. To do so, we need a clock whose ticks are of a shorter duration than that of the event so to measure events of shorter and shorter duration, we must keep subdividing the ticks. We eventually come to a limit which we take to be the Planck time at which our ticks can no longer be subdivided. A consequence is that time becomes uncertain. The same idea holds for distance and energy as well. Since neither time nor distance had any exact meaning, it follows that causality had no meaning so during this epoch the constraints imposed by causality (and entropy) did not exist.

Such inflation by itself, however, does not explain superclusters or anything else so structure is needed. We concluded that large smooth acausal vacuum structures or imprints came into existence with relative dimensions ranging from that of stars and on up to that of superclusters and beyond. Further, we show in Section 18 that the distribution of these imprints was fractal with a dimension of about 1.1. The problem then is to account for the existence of these imprints. Given the extremely short period involved and the lack of causality, it should be clear that no complicated dynamics could have been involved. Having a fractal dimension suggests that self-similarity was in some way responsible.

In **Figure 1**, we summarize the situation. The scaling is defined by $a(t) = a_P e^{\alpha(t)}$ where $\tau = \ln(t/t_P)$. The subscript "$_P$" indicates a Planck value. The exponential

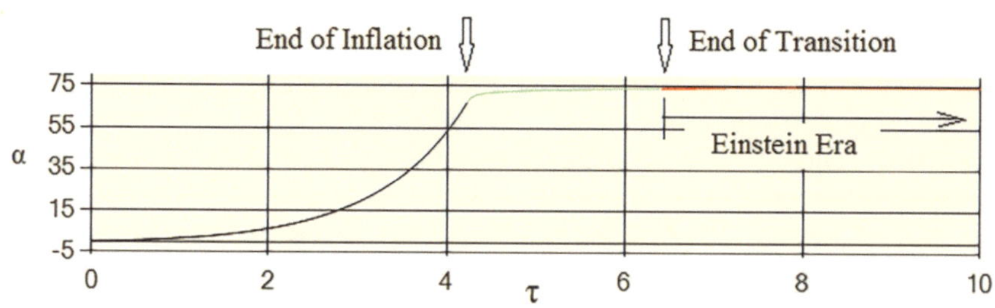

Figure 1.
Initial evolution of the universe.

expansion ended when the age of the universe became large compared to the Planck time (at about $t = 6.2 \times 10^{-42}$ s.) Following that expansion, a transition period occurred ending at a time of about $t = 6.8 \times 10^{-41}$ s. From that point onwards, the evolution of the universe could be described by Einstein's equations with the understanding that Einstein's equations have validity only for dimensions large compared to Planck dimensions. (That last statement is not an argument in support of quantum gravity which we contend does not exist. Instead, at the Planck limit, a point in spacetime does not have any precise meaning so any normal concept of dynamics cannot exist.)

As we will see in the next section, in the new solution of Einstein's equations, the scaling of the universe is entirely a consequence of the vacuum and since the model contains no free parameters, we can work backward from the present day to fix the value of the scaling at the end of the transition.

We recognize that there is no direct observational evidence that inflation existed and, in fact, there never can be for the simple reason that there was nothing to observe before nucleosynthesis. The indirect evidence for inflation, on the other hand, is overwhelming given the assumption that the Big Bang expansion began from a Planck-sized beginning. The fact that the new model solution must join onto the scaling at the end of the transition not only constrains the magnitude of the inflation, it also forces the conclusion that there was an inflation. In fact, without the inflation, the present-day size of the universe would be a small fraction of a meter.

Any discussion of a Planck era eventually brings up the point the Planck dimensions are simply dimensional constructs based on the values of certain physical constants and it would be a fantastic coincidence if the actual Planck dimensions were the same as the constructs. However, a possible resolution to this dilemma is achieved if we turn the Planck dimension formulas around giving us the origin of the physical constants, e.g. $c = l_P/t_P$ and so on.

3. A new model of the Einstein era

By the end of the transition period, the uncertainties would have become negligible and normal causality would have come into play. The development of the new model, [3], begins with the idea that the universe consists of a sequence of hyperspheres each of which is homogeneous and isotropic (the *real* universe as opposed to our preception of the universe.) The standard model goes on to assume that the curvature of each of these hyperspheres has a single, constant value but there is no physical principle that says that must be the case. We instead assert that the curvature varies with time.

We now need to connect these hyperspheres with Einstein's equations. Given the assumptions of homogenity and istropy, the hyperspheres can have no preferred origin and all their properties are dependent only on time. Our perception of the universe, on the other hand, is concerned with signals, causality, and so on and these are dependent on both time and distance as described by Einstein's equations. The question, then, is how do we reconcile the equations that describe our perception with a sequence of hyperspheres that have no notion of an origin or distance? The answer is that Einstein's equations describe the universe from the viewpoint of each observer. But a hypersphere is simply the collection of all possible observer origins so Einstein's equations become the equations that describe the hypersphere when evaluated at *any observer's origin*. Said another way, separations on a hypersphere are space-like and since Einstein's equations describe time-like events, the only possible point of contact is with a signal of zero spatial extent. Einstein's equations have nothing to say about points with spacelike separations which is the reason that Einstein's equations can accommodate multiple viewpoints of cosmology.

In the standard model, the additional assumption is made that not only are the hyperspheres homogeneous and isotropic but that the universe must *appear* homogeneous and isotropic. With time-varying curvature, on the other hand, the metric must contain an off-diagonal term connecting time with distance with the result that the universe will *not* appear homogeneous and isotropic to an observer. As we will show, however, the difference only becomes apparent at very large redshifts.

Putting these ideas together, the metric becomes,

$$\begin{aligned} ds^2 &= \left(-1+\frac{r^2h(ct,r)^2}{a(ct)^2}\left(1-k(ct)r^2\right)\right)(cdt)^2 \\ &+2h(ct,r)\,(cdt)\,rdr+a^2(ct)\left(\frac{dr^2}{\left(1-k(ct)\;r^2\right)}+r^2d\Omega^2\right). \end{aligned} \tag{1}$$

The second part of the new model concerns the vacuum energy density. Instead of the standard model concept of a vacuum described by $\mathbf{T}^{\mu\nu}=0$, the new model vacuum acts as its own source so we have

$$\mathbf{T}^{\mu\nu}=\left(\rho_{vac}c^2(ct,r)+p_{vac}(ct,r)\right)\delta^{\mu}_{0}\delta^{\nu}_{0}+p_{vac}(ct,r)g^{\mu\nu}. \tag{2}$$

After working out Einstein's equations and taking the limit as $r\to 0$, the resulting equations can be solved in closed form. The scaling is given by,

$$a(ct)=a_*\left(\frac{ct}{ct_0}\right)^{\gamma_*}e^{\frac{ct}{ct_0}c_1}, \tag{3}$$

where

$$a_*=a_0\,e^{-c_1}. \tag{4}$$

(Ref. [3] for the omitted definitions). We see that the scaling is power-law for $ct/ct_0\ll 1$ and exponential for $ct/ct_0\geq 1$. The curvature is given by

$$k(ct)=\overline{k}_0\left(\frac{a(ct)}{ct}\right)^2 \tag{5}$$

which is related to the vacuum energy density and pressure by

$$k(ct)=\frac{1}{2}\gamma_h a(ct)^2\kappa\left(\rho_{vac}c^2(ct,0)+p_{vac}(ct,0)\right). \tag{6}$$

The sum is thus a fixed function of time,

$$\rho_{vac}c^2(ct,0)+p_{vac}(ct,0)=\frac{2\overline{k}_0}{\kappa(ct_0)^2\gamma_h}\frac{(ct_0)^2}{(ct)^2}. \tag{7}$$

Note that there is no direct relationship between the scaling and the energy density or pressure. The present-day acceleration of the scaling is, in fact, a kinematic constraint that follows directly from the time variation of the curvature and the presence of the vacuum energy in the energy-momentum tensor. It has nothing to do with a cosmological constant or so-called dark energy. This situation is completely different from that of the standard model based on the FRW solution of Einstein's equations. The latter does not actually predict anything based solely on being a solution to the equations. By making choices about various parameters, it is possible to predict any sort of evolution one cares to see. In the new model that is not the case. There is one solution and *only one evolution is possible*.

Both the energy density and the pressure contain a constant term that one might think to associate with a cosmological constant. If we must assign a meaning to these constants, they are the values of the vacuum energy and pressure at infinite time. These, however, have no physical consequences and they can be removed by simply adding a constant term to the energy-momentum tensor. Physical quantities such as the curvature are functions of the sum rather than either energy or pressure individually and in the sum, the constant terms cancel.

Aside from the present-day size and age of the universe, we need the value of the scaling at two different times to fix the parameters of the model. For one, we use a present-day value of the Hubble constant. There is some uncertainty about its value but, as we will show later, a consensus of $H_0 = 73\,\text{kms}^{-1}\text{Mpc}^{-1}$ is emerging from the various methods used to determine its value. For the other parameter, we use the energy density of the CMB at the time of nucleosynthesis. From this, we determine that γ_* can be at most only slightly greater than ½. For the model development, we needed to assign a single value which we chose to be exactly ½. The relationship between the Hubble constant and the model parameters is $c_1 = t_0 H_0 - \gamma_*$ from which we determine that c_1 always has a value close to ½. For $H_0 = 73$ we have $c_1 = 0.525$.

There appears to be one remaining parameter, namely k_0. During and shortly after the inflation, the curvature was maximal which motivates an additional principle that states that *the curvature must always be as large as possible* or equivalently, that the vacuum energy density *must always be as large as possible*. From the solution, it then follows that $k_0 = 1.41$.

Everything is now fixed and unambiguous predictions can be made. We define an effective scaling parameter by $a(t) = a_0 e^{\gamma_{eff}(t)}$ which we show in **Figure 2**. We also show the scaling as a function of time. The exponential acceleration of the scaling is clearly visible.

In **Figure 3**, we show the coordinate distance of sources whose signals are received at the present plotted as a function of the lookback time. Both time-varying and constant curvature cases are shown. The two curves are similar for small values of look-back time but they differ considerably for large redshifts which illustrates the point we made earlier about the universe not appearing homogeneous and isotropic. In particular, with time-varying curvature, there is a fundamental limitation on our ability to detect distance sources. No matter how far back in time we look, we cannot see sources with coordinate distances greater than about $r = 0.6$. This is in direct contrast to the standard model where no such limitation exists.

The redshift dependence of the Hubble parameter, $H(t) \equiv \dot{a}(t)/a(t)$, is shown in **Figure 4**.

Finally, the model makes a prediction of the luminosity distances of various sources that agree with observation but we will postpone that discussion until we reach the section concerned with the Hubble tension problem.

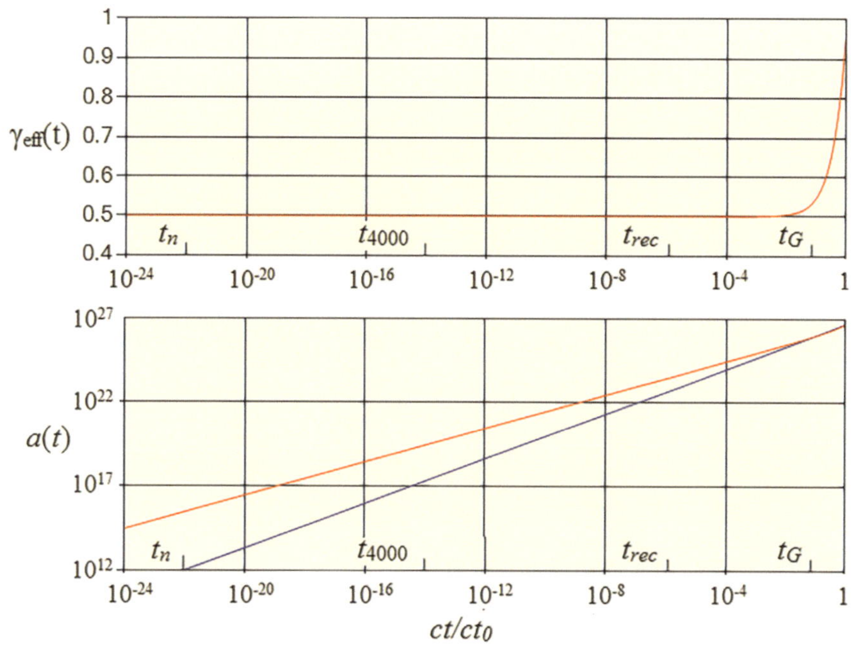

Figure 2.
The time-varying curvature predictions are shown in red. For comparison, the curve for 2/3^rds^ scaling is shown in blue. The indicated times are t_n = the time of neutron formation to be explained below, t_{4000} = the end of nucleosynthesis, t_{rec} = the time of recombination, and t_G = the time of galaxy formation.

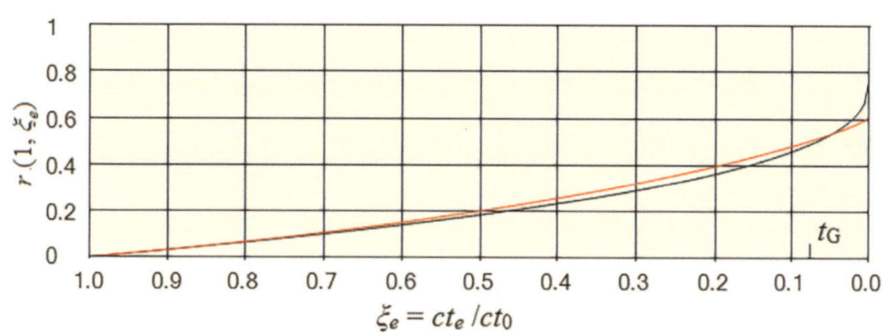

Figure 3.
$r(1, \xi_e)$ vs. ξ_e. The red curve is the time-varying curvature result. For comparison, we also show in black the result computed assuming a constant value of $k = 1$.

Up to this point, we have only considered the vacuum but eventually, particles permeated space so we need to consider their interaction with the vacuum energy density. On the LHS of Einstein's equations, the connection coefficients depend only on the metric components and these have no dependence on either the vacuum energy density or pressure. On the RHS, since the particles can be assumed to be at rest, their density is simply added to the vacuum energy density. Ordinarily, after including a new term, we would need to re-solve the equations but in this case, the equations have not changed since the particle density just becomes part of the vacuum energy sum so the original solution still holds. This means that any small variation in the particle density will be immediately canceled by the corresponding variation in the vacuum

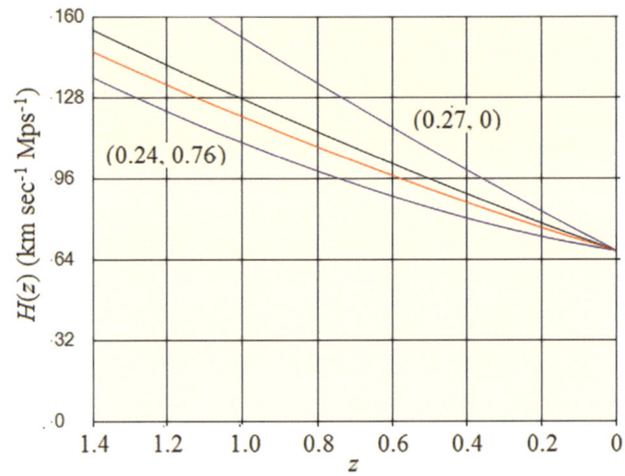

Figure 4.
Hubble parameter vs. redshift. Time-varying curvature in red, constant curvature in black. The FRW results for two values of the densities are shown in blue. The latter were normalized to the value of the new model curve at $z = 0$.

needed to keep the sum unchanged. Since the gravitational field does not change, nearby matter will not detect any difference in the field so no accretion occurs. When the vacuum energy is included, Einstein's equations show that *accretion initiated by small particle density fluctuations is impossible*.

We can now address the "coincidence" problem mentioned in the introduction. From (Eq. (7)), we find that the present-day vacuum energy density sum is

$$\rho c^2(ct_0, 0) + p(ct_0, 0) = 2.1 \times 10^{-10}\,\mathrm{jm}^{-3} \tag{8}$$

which differs from the value of the so-called dark energy density ($6.3 \times 10^{-10}\,\mathrm{jm}^{-3}$) by no more than a factor of three. Note, however, that even though the magnitudes are similar, these are in no way equivalent. The notion of dark energy driving the acceleration of the scaling just does not exist in the new model. If we assume that the average present-day particle density is about $1\,\mathrm{m}^{-3}$, we find a matter energy density equal to $E_{\mathrm{part}}(t_0) = 1.5 \times 10^{-10}\,\mathrm{jm}^{-3}$ and comparing with the vacuum value, we find that they are essentially the same. Now, using the scaling of (e.g. (3)), we determine that the particle energy density at $t = 1\,\mathrm{s}$ was $E_{\mathrm{part}}(1\mathrm{s}) = 2.1 \times 10^{17}\,\mathrm{jm}^{-3}$ and from (Eq. (7)), that the vacuum energy density was $E_{\mathrm{vac}}(1\,\mathrm{s}) = 4.0 \times 10^{25}\,\mathrm{jm}^{-3}$. The latter is larger than the particle energy density by eight orders of magnitude. Thus, what is a great problem for the standard model is a non-issue in the new model. The vast change in the energy density ratio over time is an unavoidable consequence of time-varying curvature.

4. Nucleosynthesis

The next step is to account for the creation of ordinary matter. We begin by separating what is known from what is conjecture. Observations of galaxies allow the relative abundances of the light elements to be measured. Working backward in time, the abundances at the end of nucleosynthesis can then be estimated because the processes that occurred during the intervening period are known. Similarly, the nucleosynthesis reactions are also known so one can work backward again to discover what densities of protons and neutrons were necessary to account for the abundances of the light elements. We can also establish that the process began at a time of about 10^{-5}s. That, however, is as far as one can go. Whatever happened before that time is beyond the reach of even extrapolations of observations. This means, for example, that there is no evidence to support the standard model's field theory beginning. There is also no evidence to support the standard model's assumption that the initial baryon density was uniform throughout the universe.

Here, we will propose a much simpler model, [3, 4] that leads to the same nucleosynthesis starting point but which also accounts for the matter/antimatter asymmetry of the universe, the existence of the CMB, and the high density of neutrinos needed to establish the pre- nucleosynthesis proton/neutron abundance ratio. We start with the idea that there was no existence other than the vacuum prior to the time of nucleosynthesis and that at that time, a very small percentage of the vacuum energy was converted into neutron/antineutron pairs with *a very small bias of 2–4 extra neutrons for every* 10^8 *pairs*. Using the formula for the energy density of the vacuum given earlier and assuming that the pair production occurred at the point in time when the vacuum energy density equaled the equivalent energy density of a neutron, we determine that the event occurred at a time of $t_n \approx 4 \times 10^{-5}$s.

Initially, we have neutrons and antineutrons but we also need protons. Our original idea was that weak interactions were responsible for the existence of the protons but that turned out to be wrong. The weak interactions did play a critical role but they were not responsible for the original population of protons. Instead, the protons and antiprotons were a consequence of baryonic charge exchange reactions, [4].

Starting with the neutrons and antineutrons, particle/antiparticle annihilation and charge exchange reactions proceeded at a very high rate creating, in addition to protons and antiprotons, a very high density of mesons, neutrinos and electrons along with the radiation that became the CMB. Ultimately, the small excess of initial neutrons resulted in the matter/antimatter asymmetry of the universe. This entire process was very fast and was completed 10^{-12} s after it began. At the end of this process, there were equal numbers of protons and neutrons but soon afterward, weak interactions became important and it was these that adjusted the proton/neutron ratio to the value it had when nucleosynthesis proper began.

The key point about this process is that it was not random. Instead, it was regulated by the Planck era imprint that we spoke of earlier. This imprint caused an increase relative to the average in the number of neutrons and antineutrons in the regions that became the cosmic structures and a decrease in the regions that became the voids.

For this to work, there must have been a universal violation of the baryon conservation law because the same excess had to have occurred everywhere and so cannot be the result of some random process. In fact, to account for the matter/antimatter asymmetry, some such violation must be introduced no matter what creation model is assumed. Indications of such a violation in connection with neutrons have been known for a long time. The evidence, known as the neutron enigma, comes from experiments conducted to determine the half-life of the neutron. The problem is that different values result depending on the design of the experiment, (see reference in [3]). Experiments that detect the protons that result from the neutron decay give one value and experiments that count the total number of neutrons remaining in a container as a function of time gives another value. The difference is evidence that some of the neutrons simply disappear without leaving behind any daughter particles.

Referring now to the problems listed in [1], there are several that can now be settled. There are several supposed anomalies with the CMB spectrum that are again artifacts of the ΛCDM model. Problems such as the CMB cold spot, cosmic hemispherical power asymmetry, the lack of large-angle CMB temperature correlations, the cosmic dipole problem, and the galaxy cluster count problem among others do not exist in a universe in which the creation of matter (and radiation) was regulated by a fractal vacuum imprint. The universe began with structure instead of with a uniform matter distribution.

We will discuss this initial stage of nucleosynthesis more fully in a later section but to appreciate the details of this process, we first need to examine the subsequent development of cosmic structures.

5. The origin of cosmic structures

Skipping over the details of nucleosynthesis proper, we will now consider the problem of how the proton gas that resulted from nucleosynthesis became the structures we now observe. The filament structure that defines superclusters was fixed by the nucleosynthesis process discussed above and because the filaments are vastly too large to undergo gravitational evolution, they are much the same today as they were initially. The components that make up the filaments such as galaxy clusters, on the other hand, did evolve.

To study the initial evolution of all smaller cosmic structures, we consider the motion of a particle lying at the outer edge of a volume containing the mass that eventually became the structure in question, [5]. The only accelerations acting on the particle were gravitation and the acceleration due to the expansion of the universe so the equation of motion of the particle becomes

$$\begin{aligned}\ddot{R}(t) &= -\frac{GM_{eff}(t)}{R(t)^2} + R_0 f_s \frac{\ddot{a}(t)}{a_0} \\ &= -\frac{GM_{eff}(t)}{R(t)^2} + R_0 f_s \left(-\gamma_* + \left(\gamma_* + c_1 \frac{t}{t_0}\right)^2\right) \frac{a(t)}{a_0} \frac{1}{t^2}.\end{aligned} \tag{9}$$

The coordinate $R(t)$ is the distance from the particle to the center of the structure and

$$M_{eff}(t) = M_{Struct}\left(1 - \left(\frac{R(t)}{R_{Sur,s}(t)}\right)^3\right) \tag{10}$$

is the effective mass of the structure adjusted for the presence of the background particle density.

The coordinates are illustrated in **Figure 5**. R_0 is the present-day size of the structure. The parameter f is a multiplier used to define the size of any feature of interest in present-day terms. The size of the same feature at any earlier time is given by $R(t) = R_0\,(a(t)/a_0)f$. The black lines indicate the change in the distance due to the expansion alone. The lower black line indicates the size that would have evolved into

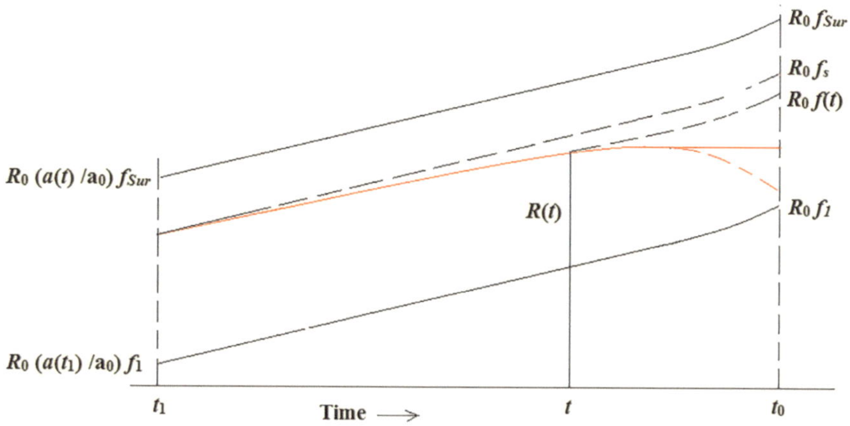

Figure 5.
Coordinate system for structure evolution.

the present-day size in the absence of gravitational acceleration (the value at $t = t_0$ is its present-day size.) The red line illustrates a possible trajectory for a particle when gravitation is included.

If we consider a particle located at the radius of the volume needed to form any structure from the background density, nothing would happen because the density would be the same everywhere. The corresponding multiplier is f_{sur}. To get things going, we need to fix the starting radius of the volume containing the mass of the structure at a value less than the value corresponding to the background. The multiplier in this case is $f_s < f_{sur}$.

We will first consider the evolution of a galaxy cluster. In all cases, we imagine that the particles were created at rest so their initial recession velocity was fixed by the expansion. According to the accretion model of structure formation, structures could not come into existence until there was something to accrete which we take to mean that the starting time would be the generally accepted time of galaxy formation. In **Figure 6**, we show the model results for several values of initial over-density. The blue line is the solution for $f_s = f_{sur}$ and since in that case there was no gravitational acceleration, the size would follow the expansion of the universe. Considering the red lines, it is apparent that a value of $f_s < 7$ is necessary for the cluster to have any chance of forming, and a value closer to 4 would be necessary to account for the present-day size. The point at which the structure ceases to expand we call the zero-velocity point (ZVP). The critical point is that the cluster must have already been a well-defined structure long before the time conventionally associated with galaxy formation.

Since we have argued earlier that all structures came into existence at the time of nucleosynthesis with more or less their final mass, we recalculate the evolution beginning at the time that nucleosynthesis proper ended or about $t_s = 4000$ s. The procedure was to run simulations with decreasing values of the initial radius but enclosing the same total mass until we obtained a structure with its present-day size. The results are shown in **Figures 7** and **8**. The red line shows the evolution of the cluster and we see that it reaches its ZVP at $t = t_G$ and with its present-day size. The other curves show the paths of a few internal and external galaxies. The galaxy with $f_s = 12.4$ is an example of an external galaxy that was too far from the cluster to be captured.

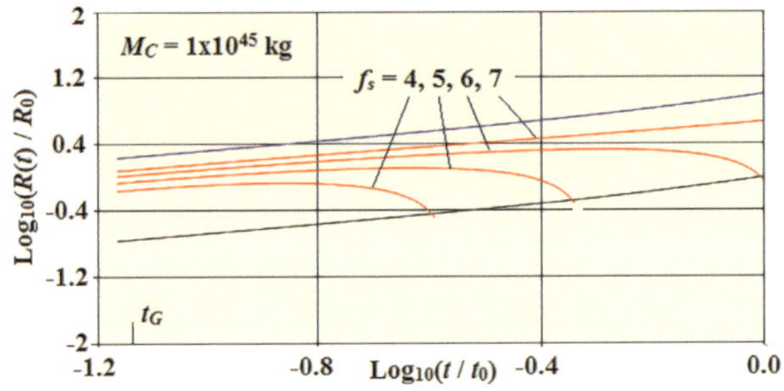

Figure 6.
Solution for four values of the starting position with $t_s = t_G = 3.2 \times 10^{16}$ s *and a cluster mass of* $M_C = 10^{45}$ kg.

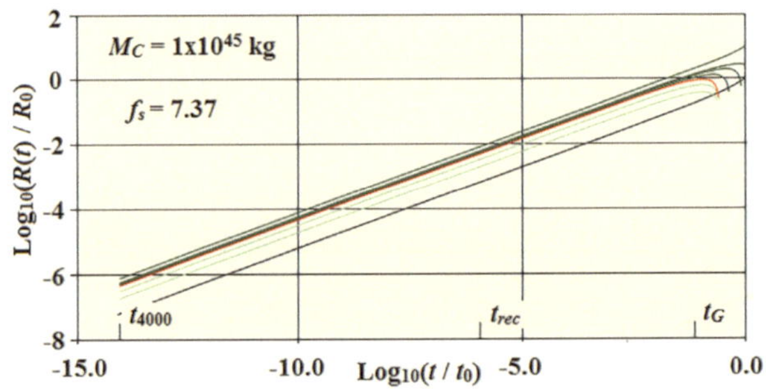

Figure 7.
Solution with $t_{4000} = 4000$ s *and* $f_s = 7.37$.

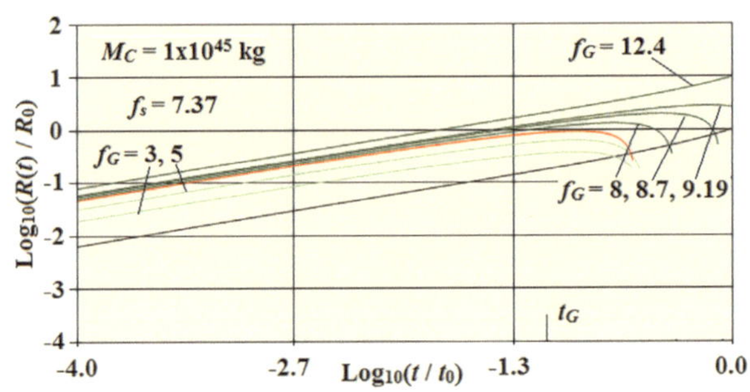

Figure 8.
Detail view of ***Figure 7***.

The value $f_s = 7.37$ corresponds to an initial compaction ratio of 1.35 or an initial density of 2.5m^{-3} in present-day terms. What is apparent is that the motion of the surface was completely dominated by the expansion up until a time fairly close to $t = t_G$. Until then, the particles were moving away from each other far too rapidly for gravitation to have any effect on their evolution. The recession rate, $\dot{R}(t) = R_0(\dot{a}(t)/a_0)f$, for example, does not drop below the speed of light until a time of about 10^{10} s. Keep in mind that during this phase, the particles are not actually moving. Eventually, however, gravitation does begin to slow the expansion, and finally, at the time usually associated with galaxy formation, the expansion stops at the ZVP. From that point, the equations indicate that the cluster would have undergone free-fall collapse which it would have done in the absence of processes that stabilized the cluster. An interesting point about clusters is that the evolution shown for the Virgo cluster seems to apply generally to all clusters. They reach their ZVPs at a time of about $t = t_G$ with their final present-day sizes over a considerable range of masses.

6. Galaxy evolution

We will now apply the same model to galaxies, [5]. The procedure is the same as for clusters. Taking the Milky Way as our example, the outer surface parameter is $f_{sur} = 140$. The results for three values of f_s are shown in **Figure 9**.

A value of $f_s = 55$ results in a ZVP close to $t = t_G$. This corresponds to an initial proton density of about $17\,\text{m}^{-3}$ in present-day terms and an initial size about 55 times its present-day size again in present-day terms. The really interesting prediction of the model, however, is that galaxies reached their ZVPs with sizes many times larger than their present-day sizes. For the Milky Way, the ratio was 5.2. This result turns out to be of critical importance. It is responsible for the existence of the observed large HI rings, [6], and as we will show below, for the existence of supermassive black holes which, in turn, were responsible for the radiation that stabilized those same galaxies and also galaxy clusters, [7].

At this point, we emphasize that in this new model of structure formation, all structures came into existence at roughly the same time with essentially all their present-day mass. Clusters had most of their complement of galaxies, and galaxies had a good supply of stars. This result is completely at odds with what has been expected based on the accretion model of galaxy formation. In the latter model, the earliest galaxies would have be fragmentary rather than fully formed. *The early results from the James Webb telescope clearly support the new model viewpoint over that of the accretion model.*

7. Stellar evolution

We will next move on to the stars, [5]. Although stars are extremely dense compared to everything else, that density is a consequence of their collapse. At the time of

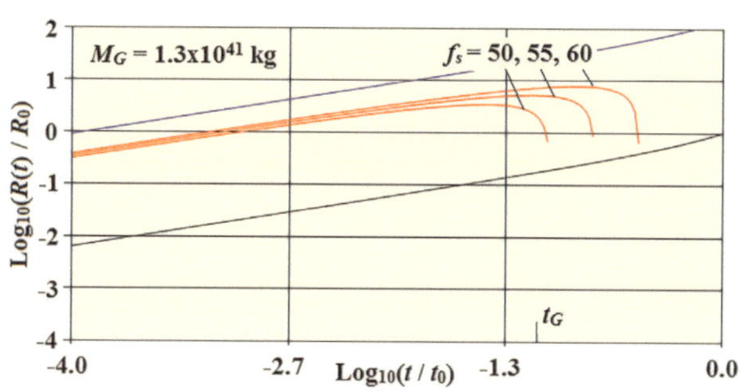

Figure 9.
Galaxy evolution for three values of f_s and a mass of $M_G = 1.3 \times 10^{41}$kg.

nucleosynthesis, their imprint density was only slightly larger than that of their host galaxies. For our model, we assume an initial nominal dimension of 1 ly in present-day terms. It turns out that with stars, gravitation became effective much earlier in their evolution than was the case for the larger structures. Depending on the initial value of f_s, stars could reach their ZVP at times even earlier than $t = t_{rec}$ which would imply that we could have had stars long before the formation of the galaxies or even recombination. The fact that they did not undergo early collapse is a consequence of the temperature of the gas.

According to Jean's model of star formation, for a star to form, the sum of the potential and kinetic energies must be less than zero which results in a critical temperature of

$$T_c = \frac{0.683}{R(ly)}\ \mathrm{K} \tag{11}$$

where R is the radius of the cloud at the ZVP. Before t_{rec}, the gas was in thermal equilibrium with the CMB radiation. At t_{rec}, the gas had a temperature of about 3000 K, and thereafter, the temperature of the gas decreased proportionally to $a(t)^{-2}$. In **Figure 10**, we compare the actual temperature with the critical temperature.

We find that the gas temperature did not reach the critical value until about $t = t_G$ when it had a value very close to 1 K. The result was that stars condensed into their final form at the same time as the galaxies but for a different reason.

8. Galaxy and Galaxty cluster stability

We have now established that, aside from superclusters, all structures reached their ZVPs at approximately the same time and with essentially all their present-day mass. The problem is that, without some process to prevent it, both galaxies and galaxy clusters would have undergone the gravitational free-fall collapse indicated in the figures. (The stars did undergo collapse so we are not concerned about them.). The reason that galaxies and clusters did not collapse was that their constituent gas was rapidly heated very soon after they reached their ZVPs, [7].

To model the dynamics, we used a modified form of the usual hydrodynamic equations. We assumed spherical symmetry and adopted the Lagrangian viewpoint. The final equations in terms of dimensionless coordinates are as follows;

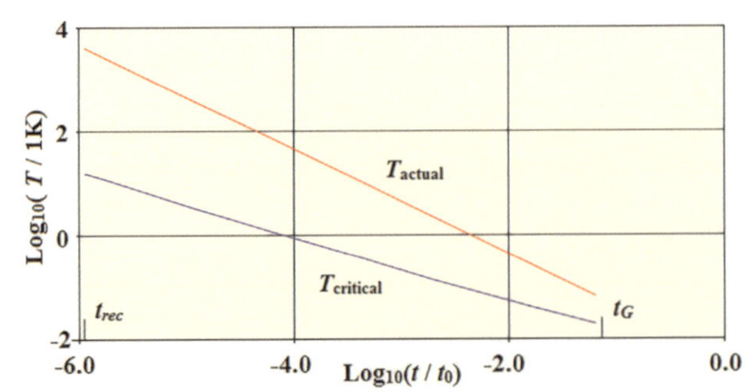

Figure 10.
Comparison of the actual and critical temperatures.

$$\frac{\partial \overline{r}}{\partial \overline{t}} = \overline{v} \tag{12}$$

$$\frac{\partial \overline{v}}{\partial \overline{t}} = -\frac{\overline{m}}{\overline{r}^2} - 4\pi \frac{\tilde{\rho}^{5/3}}{\overline{r}^{4/3}} \left(\frac{\partial \overline{\psi}}{\partial \overline{m}} + \frac{5}{3} \frac{\overline{\psi}}{\tilde{\rho}} \frac{\partial \tilde{\rho}}{\partial \overline{m}} \right) + \frac{10}{3} \frac{\tilde{\rho}^{2/3} \overline{\psi}}{\overline{r}^{7/3}} + \frac{\overline{r}}{\overline{t}^2} \left(-\gamma_* + \left(\gamma_* + c_1 \frac{\overline{t}}{\overline{t}_0} \right)^2 \right) \tag{13}$$

$$\frac{\partial \tilde{\rho}}{\partial \overline{t}} = -4\pi \tilde{\rho}^2 \frac{\partial \overline{v}}{\partial \overline{m}} \tag{14}$$

$$\frac{\partial \overline{\psi}}{\partial \overline{t}} = \frac{2}{3} \left(\frac{\overline{r}^2}{\tilde{\rho}} \right)^{2/3} \overline{q}. \tag{15}$$

We have defined a dimensionless time, $\overline{t}$, by

$$t = t_s(0.469 + \overline{t}) \tag{16}$$

where $t_s = \sqrt{R_c{}^3/GM_C} = 6.8x10^{16}$s so $\overline{t}$ ranges from 0 to 5.91 as t ranges from t_G to t_0. We have eliminated the pressure in favor of the entropy function,

$$\overline{\psi}(\overline{t}, \overline{m}) = \frac{p(\overline{t}, \overline{m})}{\rho(\overline{t}, \overline{m})^{5/3}} \tag{17}$$

and have defined an adjusted density by

$$\tilde{\rho}(\overline{t}, \overline{m}) \equiv \overline{\rho}(\overline{t}, \overline{m})\, \overline{r}^2. \tag{18}$$

The independent spatial coordinate, $\overline{m}$, is related to the radial coordinate by

$$\overline{m} = 4\pi \int_0^{\overline{r}} d\overline{r}'\, \tilde{\rho}(\overline{t}, \overline{r}'). \tag{19}$$

We note that these equations do not contain any dimensionless structure-dependent parameters so the solutions apply to both galaxies and clusters. Initially, the gas had a temperature of at most a few K, and because we are starting at the ZVP, $\overline{v}(\overline{t}, \overline{m}) = 0$. We also know their initial size and mass.

We will first consider the evolution of clusters. To establish the time scale of our problem, in **Figure 11** we show the evolution of a cluster in the absence of any heating and find that the stabilization of the clusters must occur very soon after the ZVP.

We have two problems to solve. The first is to determine what combinations of the initial density and radiation result in a cluster that neither collapses nor evaporates. The second problem is to identify the source of the radiation. We soon discovered that with a uniform initial density profile, clusters would have collapsed no matter what temperature profile is assumed. The same is also true if the profile is too sharply peaked at the origin. We tried several initial density profiles and determined that the profile must have a moderate negative slope. An example is the curve labeled $\overline{\rho}(\overline{r})$ in **Figure 12**.

We next consider the heating. We very quickly determined that compressive heating develops far too slowly to prevent the collapse which means that a source of radiation must have been responsible. We thus needed to assume a radiation source profile and our first try was the broad profile shown in **Figure 13**. The sample profile

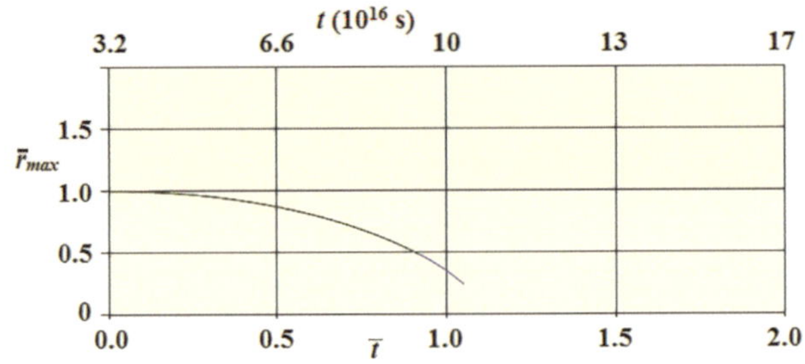

Figure 11.
Cluster free-fall collapse.

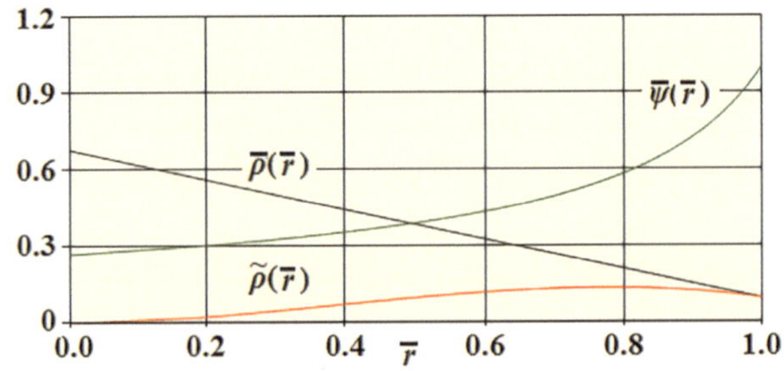

Figure 12.
Linear density profile with negative slope.

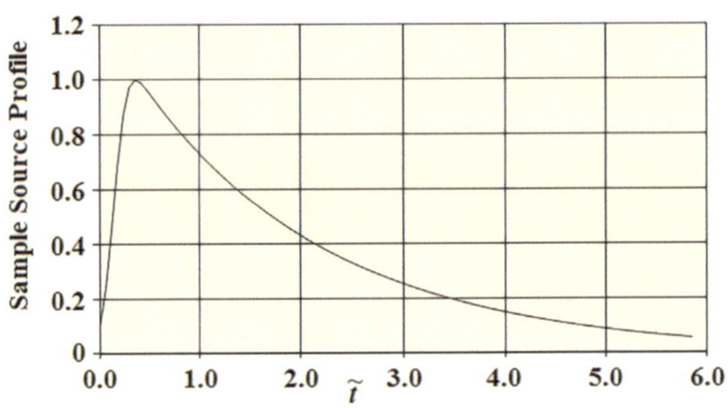

Figure 13.
Broad radiation profile.

has a maximum value of unity which is then scaled by a multiplier in the simulation code. With a multiplier of five, the resulting size of the cluster is shown in **Figure 14**. What we find is that while a significant amount of radiation is necessary to prevent the collapse, it must also be short-lived because otherwise, the cluster would have evaporated.

For our next try, we assumed the narrow profile shown in **Figure 15** and obtained the evolution shown in **Figure 16**. (Narrow is a relative term, the width of the peak still amounts to $\approx 10^9$yr.) This result shows that we are on the right track. We have found a solution that neither collapses nor exhibits undue expansion. The necessary conditions are that the density profile must have a modest negative slope and that the

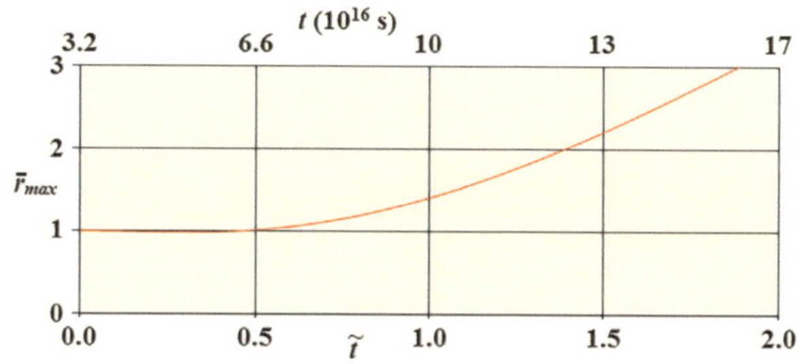

Figure 14.
Solution with the density and radiation profiles of ***Figures 12*** *and* ***13****.*

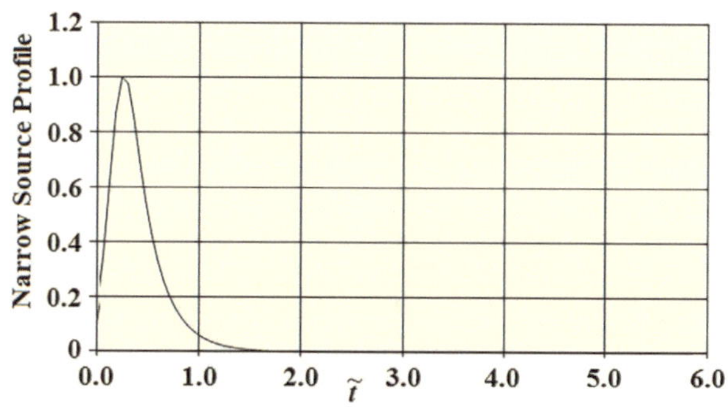

Figure 15.
Narrow radiation profile.

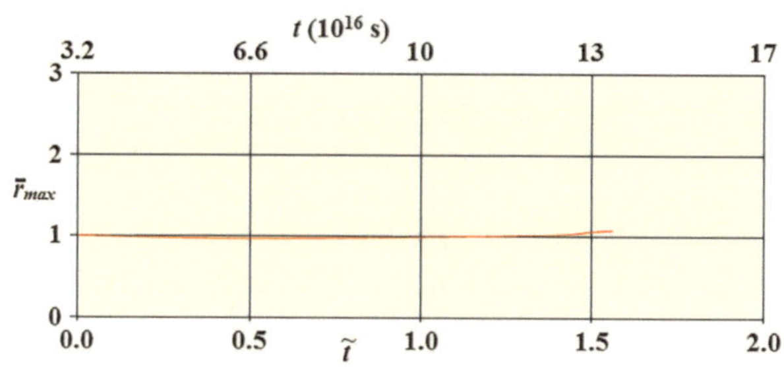

Figure 16.
Solution with narrow radiation profile and a multiplier of 6.

radiation profile must rise sharply after the ZVP and then, just as quickly it must cease or nearly cease.

The actual peak radiation intensity is given by $q_{peak} = f_m 1.4x10^{-5} \text{jkg}^{-1}\text{s}^{-1}$ where f_m is on the order of 5 or 6. Multiplying by the mass of the cluster, the total absorbed radiation works out to be on the order of $f10^{40}$watts or $f2.5(\text{M}_{\odot}\text{c}^2)\,\text{yr}^{-1}$.

The final problem is to identify the source of this radiation. We need a huge power output and even more importantly, a significant lifespan and active galactic cores, or, in other words, quasars are the only real possibility. Power outputs from large quasars are on the order of 10^{40}watts which is in the range needed. A few large quasars, however, would not solve our problem. First, the released radiation would be too

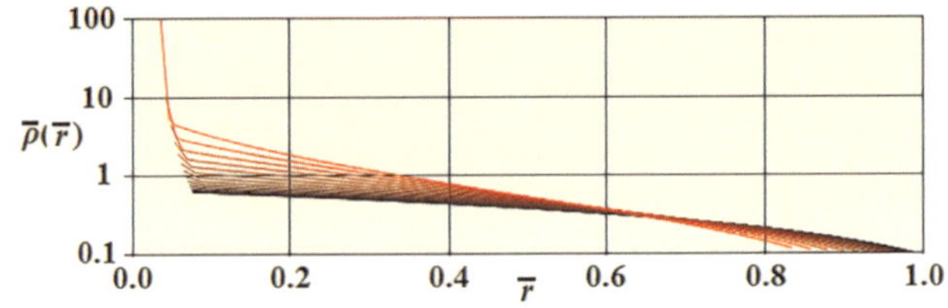

Figure 17.
Density profiles during the initial phase of galaxy collapse.

concentrated in some areas and insufficient in others, and second, it would have lasted too long. Instead, we need a large number of short-lived mini-quasars which would have run out of fuel in a relatively short time and become normal galaxies.

To have quasars, however, we must have supermassive black holes. We discovered earlier that galaxies would have undergone gravitational free-fall collapse soon after reaching their ZVP unless something prevented that from happening. The galaxies reached their ZVP with sizes several times larger than their present-day size and soon after did begin to collapse. Initially, the gas making up the galaxy was cold and, as was the case of the clusters, compressive heating would not have stopped the collapse. We now assert that during this contraction, all galaxies formed supermassive black holes together with accretion disks. In **Figure 17**, we show a calculated sequence of density profiles showing the initial phase of the collapse. We find that the density increases extremely rapidly at the origin creating the conditions necessary for the formation of a black hole and that it does so long before there is any significant reduction in the radius of the galaxy.

As the collapse continued, an accretion disk must have formed and the resulting radiation heated the galactic gas thereby stopping the collapse. The key point is that *all galaxies must have formed a supermassive black hole complete with an accretion disk* because they would have otherwise collapsed.

Much of that radiation would have escaped the galaxy and in the cases in which the galaxies were located inside a cluster, they would have heated the cluster gas. The Virgo cluster contains a large number of galaxies, most of which are dwarf ellipticals. References listed in [7] found that first, quasar host galaxies are all ETGs with the bulk being ellipticals and second, that dwarf ellipticals do have active nuclei. The Virgo cluster does not contain a quasar at present but it is a characteristic of active nuclei that they radiate huge amounts of energy until their supply of accretion material runs out after which they become normal galaxies. The conclusion is that clusters were heated from within by a large number of mini-quasars that collectively used up their accretion supplies within the time scale indicated in **Figure 15**.

9. Nucleosynthesis again

We will now return to the nucleosynthesis pair production process that we introduced earlier. To study the evolution, we created a simulation model that included neutrons, protons, electrons, neutrinos, photons, and their antiparticles. We then solved the coupled reaction rate equations to determine the evolution, [4].

As we discussed earlier, at a time of about $t_n = 4 \times 10^{-5}$ s a very small percentage of the vacuum energy was converted into neutron and antineutron pairs with a neutron excess of about two extra neutrons for every 10^8 pairs. The list of reactions is long but the principal ones were the baryon annihilation and charge exchange reactions listed below.

$$\mathrm{n} + \overline{\mathrm{n}} \rightarrow \gamma + \text{leptons} \tag{20}$$

$$\mathrm{p} + \overline{\mathrm{p}} \rightarrow \gamma + \text{leptons} \tag{21}$$

$$\mathrm{n} + \overline{\mathrm{p}} \rightarrow \gamma + \text{leptons} \tag{22}$$

$$\overline{\mathrm{n}} + \mathrm{p} \rightarrow \gamma + \text{leptons} \tag{23}$$

$$\overline{\mathrm{p}} + \mathrm{p} \rightleftarrows \mathrm{n} + \overline{\mathrm{n}}. \tag{24}$$

(We have indicated the final states of the annihilation reactions but these actually pass through an intermediate meson state that eventually decays into the photons and leptons.) In **Figure 18**, we show the results for an initial density of $n_{\mathrm{n}}(0) = 10^{41}\mathrm{m}^{-3}$ and a neutron excess ratio of 2×10^{-8}. The whole process was completed in about 10^{-12} s.

The simulation was primarily concerned with the baryon annihilations and the baryon curves are correct. The details of the product particle distributions, on the other hand, were of only secondary interest at the time and because we needed to keep the simulation manageable, we did not take proper account of the meson state branching ratios or the meson lifetimes which is the reason that the leptons and photon curves are shown dashed.

The annihilation decay products are primarily pions. Neutral pions decay into two photons with a relatively short mean lifetime so labeling the curve as photons is a fair representation of the actual situation. Charged pions, on the other hand, decay into muons which subsequently decay into either electrons or positrons plus neutrinos with a combined mean lifetime much larger than the time scale of the figure. That being the case, the curves labeled $\mathrm{e}^+, \mathrm{e}^-$ and $\nu, \overline{\nu}$ could more accurately be labeled π^+, π^- but since each charged pion eventually decays into either a single electron or positron plus neutrinos, it is fair to label the densities as those of the leptons since those are approximately the densities they would eventually have.

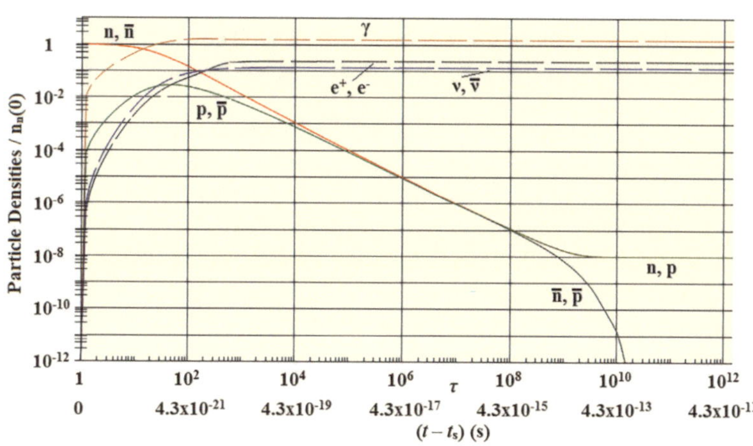

Figure 18.
Initial stage of nucleosynthesis.

One of our assumptions was that this process began at the point when the vacuum energy density equaled the equivalent mass energy of a neutron. Using a value of 0.5×10^{-15}m for the radius of the neutron, we find that the vacuum energy was equivalent to a neutron number density of $2 \times 10^{45}\text{m}^{-3}$ so that the initial density chosen corresponds to a fraction of only 5×10^{-5} of the total vacuum energy. The combination of the charge exchange reactions and annihilations rapidly eliminated the antibaryons leaving a final baryon density of $n_{bary} = 2 \times 10^{33}\text{m}^{-3}$ which we compare with a value $4 \times 10^{33}\text{m}^{-3}$ obtained by scaling a present-day density of 1m^{-3}. This total density thus represents a fraction of the total vacuum energy of only 10^{-12} so that, of the very small percentage of the vacuum energy that was converted into matter, only an extremely small percentage ended up as baryonic matter.

The number density of photons produced was on the order of $n_\gamma \sim 1.5 \times 10^{41}\text{m}^{-3}$. The radiation had not yet reached a black-body spectrum within the period illustrated but the total energy of the created photons corresponds to a black-body temperature equal to $T_\gamma = 2.7 \times 10^{11}$K which we compare with $T_{CMB} = 4.3 \times 10^{11}$K obtained by scaling the present-day CMB temperature back to the time of this event.

The number density of neutrinos/antineutrinos was also large, $n_{\nu,\bar{\nu}} \sim 1.2 \times 10^{40}\text{m}^{-3}$. This result is a major success of this model because it explains the origin of the high density of neutrinos without which the weak interaction rates would not have been great enough to adjust the proton/neutron density ratio to the value necessary for nucleosynthesis proper.

10. CMB anisotropy

We have just seen that the model can account for the existence of radiation with a temperature corresponding to that of the CMB. To understand the anisotropy, we recall that departures of the matter densities from the background value were necessary to account for the development of all cosmic structures. For example, the Virgo cluster density excess was 1.35 and that of the Milky Way was 17. To achieve those excesses, the vacuum pair production rate must have been corresponding larger by the same factor. Conversely, in the void regions, it must have been smaller. But the number of produced photons varies in direct proportion to the initial particle density as does the total energy of the radiation. Since the temperature of black-body radiation varies as T^4, doubling the initial density of neutrons would increase the temperature by about 20%.

We find then that *the anisotropy of the CMB was built in at the time the radiation was created*. As the superclusters were created, so were the peaks and valleys of the CMB spectrum. This is another major departure from the standard model viewpoint.

Referring back to the previous section, the simulation was carried forward to the point that the baryonic annihilations were complete. Large numbers of photons, electrons, and neutrinos were then in existence, and the counts of the protons and neutrons were equal. The next step is to carry the evolution forward to the start of nucleosynthesis proper. During this period, the spectra of the radiation and particles became black body, the proton/neutron ratio was adjusted to a value of 80/20 by the weak interactions. At the end of the annihilation phase, while the reaction rates between the just created particles, photons, and neutrinos were very high, they were all in equilibrium so no net energy was being transferred. The radiation spectrum, for example, was static. What is missing in the simulation are reactions that transfer energy away from the protons because until those energies are much lower, the weak interactions remained symmetric and no adjustment of the proton/neutron ratio could occur.

To model this phase of the evolution, we must first take into account the annihilation meson branching ratios and the meson mean lifetimes mentioned earlier and then include the known deep inelastic baryon reactions in which one or more mesons are produced. An example is the set of reactions $p + p \rightarrow p + p +$ mesons. Such reactions were not an important factor during the annihilation phase but they played a critical role immediately afterward because the mesons in such reactions carry away a significant amount of energy. Our work on this simulation is not complete at this point but we expect to finish it during the next few months. The simulation has advanced to the point that we can say that the CMB radiation reaches a black-body spectrum almost immediately and very soon after, the protons, neutrons, pions, and electrons reach thermal equilibirum with the radiation with a black-body rather than Maxwell-Boltzmann.

11. The connection between the CMB and superclusters

Figure 19 shows the well-known CMB anisotropy map. A portion of the map has been enlarged in the lower rectangle and two angular size references are also included.

The CMB we receive was emitted by a spherical shell whose radius is fixed by the coordinate distance of **Figure 3** when evaluated at the time of recombination, [3]. Thus, $S(t_{rec}) = 0.6a(t_{rec})$. For a structure of size, $D(t_{rec})$, the subtended angle would then be

$$\theta = \frac{D(t_{rec})}{S(t_{rec})}(360/2\pi). \tag{25}$$

An important property of the vacuum metric is that there are no off-diagonal components linking time and the angular coordinates which means that the angle between two approaching signals does not change. This being the case, we can move this equation forward to the present day,

$$\theta = \frac{D(t_0)}{0.6a(t_{rec})}\frac{a(t_{rec})}{a(t_0)}(360/2\pi) = 95.5\frac{D(t_0)}{a(t_0)}\text{deg}. \tag{26}$$

In **Figure 20**, we see that the first peak of the CMB spectrum has an angular size of about 1 deg.

In **Table 1**, we show the angular sizes of various structures and we see that superclusters (and voids) are the only structures large enough to be associated with the peak.

In **Figure 21**, we show a plot of 71 known superclusters and voids. Using the Gaussian distribution shown, we performed a statistical calculation to determine the spectrum with the result shown in **Figure 22**.

We find that the position of the peak is correct. The shape of the predicted peak is slightly broader than the observed peak but that is quite likely due to our assumption of spherical symmetry which is certainly not the case. The magnitude of the peak was adjusted to match the observed peak and is yet not a prediction but we do expect that a prediction will result from our new simulation mentioned previously that is still

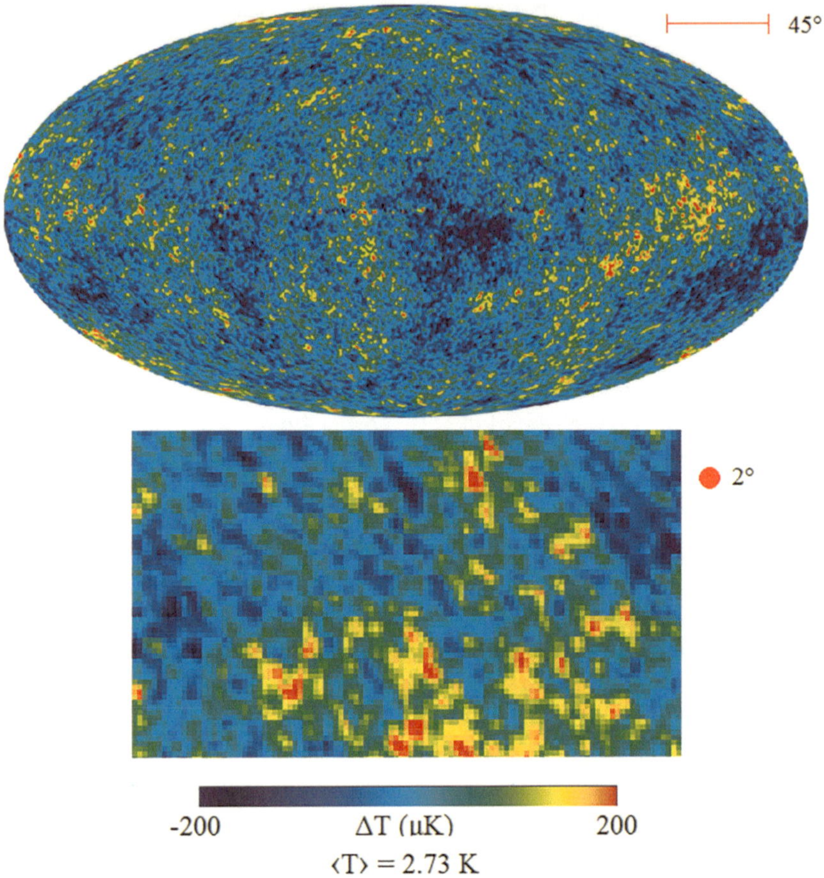

Figure 19.
CMB anisotropy.

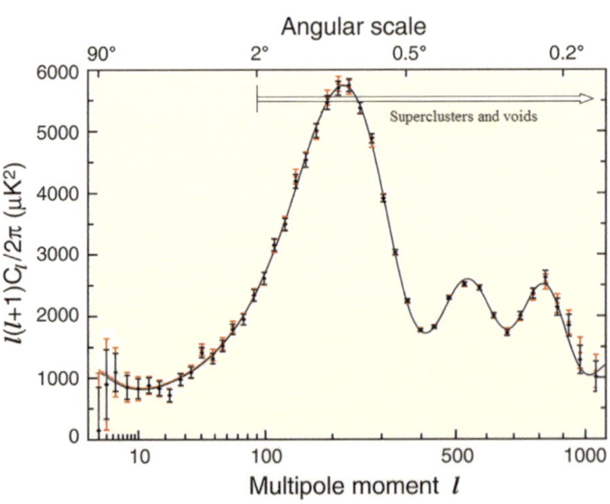

Figure 20.
CMB spectrum.

Object	θ (deg)
Milky Way	0.0001
Groups	0.007–0.013
Clusters	0.013–0.065
Superclusters	0.2–2.0
voids	0.6–1.6
Extreme structures	> 45

Table 1.
Angular sizes of various structures.

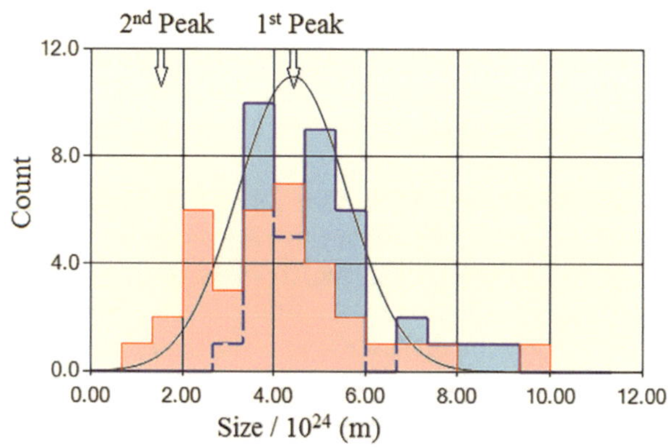

Figure 21.
Count of observed superclusters (red) and voids (blue).

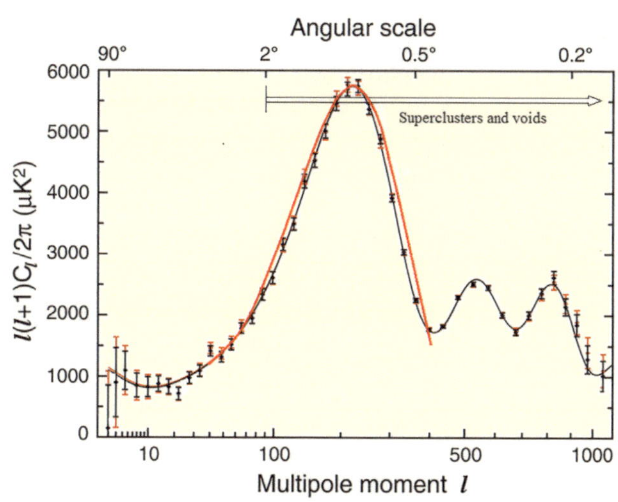

Figure 22.
Ensemble average supercluster/void CMB spectrum. Angles are related to the moment by $l = \pi/\theta_{\text{rad}} = 180/\theta_{\text{deg}}$.

under development The second peak does not correlate with the size of any structure which is strong evidence that it is a consequence of multipole distributions within the superclusters and voids. Referring back to **Figure 19**, we see that the supercluster-sized structures have a range of temperatures which supports that idea.

We remind the reader that even though the superclusters have the right size, it wasn't the superclusters per se that were responsible for the peak but rather that the process that created the superclusters simultaneously created the CMB spectrum peaks with the same dimensions.

12. Baryonic acoustic oscillations: CMB

Since this bears on the CMB anisotropy problem, this is an appropriate point to show that the baryonic acoustic oscillation model is wrong, [8]. We will first discuss the CMB issue and then the unrelated 2-point galaxy correlations model. One of the principal failings of the BAO model is that the authors assumed from the beginning that there existed a uniform distribution of radiation that needed a spectrum. We have shown, however, that that idea is simply wrong. The radiation came into existence with its spectrum in place from the beginning.

The first argument against the BAO model is simply causality. In **Figure 23**, we show the coordinate of a photon emitted at the time of nucleosynthesis in comparison with the coordinate size of a supercluster which we have shown has the same dimension as the spectrum peak. We see that by the time of recombination, the photon could have only traveled about 10% the distance across the supercluster.

The second argument is concerned with the energy required to create the peaks starting from an energy source that emitted acoustic waves at the time of nucleosynthesis. At present, the relative magnitude of the CMB anisotropy is $\Delta T_0/T_0 = 2 \times 10^{-4}$ and because the CMB decoupled from matter in the universe at the time of recombination, the ratio had the same value earlier at $t = t_{rec}$ when the temperature was 3000 K. The energy density of the anisotropy was $\rho_\Delta = a_{SB}\left((T + \Delta T)^4 - T^4\right) = a_{SB}T^4(\Delta T/T)$ which gives $\rho_\Delta = 4.9 \times 10^{-5}\text{jm}^{-3}$.

Also, at that time the average size of a supercluster was 4×10^{21}m so a spherical shell, with a radius of half the supercluster size and an assumed thickness equal to 10% of the radius, had a volume of $V \sim 10^{64}\text{m}^3$. Multiplying gives a total energy of 4.9×10^{59}j. The BAO model assumes that this energy originated from a concentrated source whose size at the time of nucleosynthesis could not have been larger than ct_n.

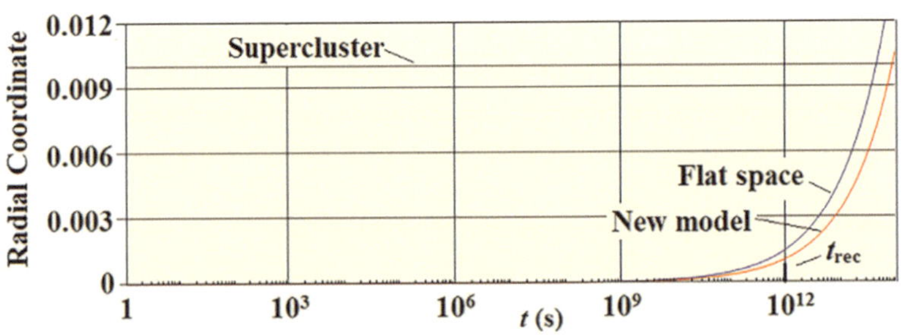

Figure 23.
BAO causality restriction.

The required energy density would then have been $\rho_{\text{BAO}} = 1.2 \times 10^{47} \text{j m}^{-3}$ which is 3×10^{12} times greater than the total energy density of the universe. This argument is oversimplified but 10^{12} is a big number so it is not likely that a more sophisticated model will change the result. If one tries account for the energy by proposing a huge number of sources, one is immediately faced with the problem of specifying how to arrange, along with what did the arranging, the sources in space so that they produced the cosmic web and not some uniform distribution.

The conclusion is that *baryonic acoustic oscillations on a scale needed explain the CMB spectrum are impossible*.

13. Baryonic acoustic oscillations: Galaxy correlation function

The second phenomenon attributed to BAO is the bump observed in the two-particle galaxy correlation function at a separation of 150 Mpc, [8]. The existence of that bump would present a problem if BAO were the only possibility since we have argued that BAOs do not exist. There is, however, another and far simpler explanation, namely the cosmic web.

To test this idea, we built a simulation based on a cubic grid with edge lengths initially set to the average size of a supercluster. Using Gaussian statistics, we randomized the ends of the edges both in magnitude and direction according to the size distribution of superclusters. Next, we assigned a random number of galaxies to each edge with random offsets scaled by the width of a supercluster and finally, calculated the correlation function of the resulting distribution. The set of galaxies along the center grid line formed our reference set. There are a number of estimators described in the literature. The one we used is the spherically symmetric estimator given by the following (see ref. [8]),

$$1 + \xi(r) = \frac{1}{N}\sum_{i=1}^{N} \frac{N_i(r)}{\overline{n}\, V_i(r)} \tag{27}$$

In this formula, the number of reference galaxies is N and $\overline{n}$ is the average density of all the galaxies. $N_i(r)$ is the number of galaxies lying within a thin spherical shell of radius r and volume $V_i(r)$ centered on the ith reference galaxy. This ratio compares the actual number in each shell with the number expected based on the average density of galaxies. If the galaxies were distributed randomly, we would have $\xi(r) = 0$.

The result of running the simulation for a number of different random number sets is shown in **Figure 24**. While the model is perhaps not an accurate representation of the cosmic web, the result looks very much like the observed correlation function in shape and the peak appears at the correct distance.

During the past two decades, a lot of effort has been spent on the galaxy correlation problem because of its potential as a standard candle for measuring distances at redshifts beyond the supernovae scale. What we have just shown is that the idea is perfectly reasonable because the coordinate dimensions of the cosmic web do not change. In all the work that has been done, one only needs to replace the wording" baryonic acoustic oscillation" with" cosmic web."

Taken together, we have shown that *BAOs are neither possible nor necessary*.

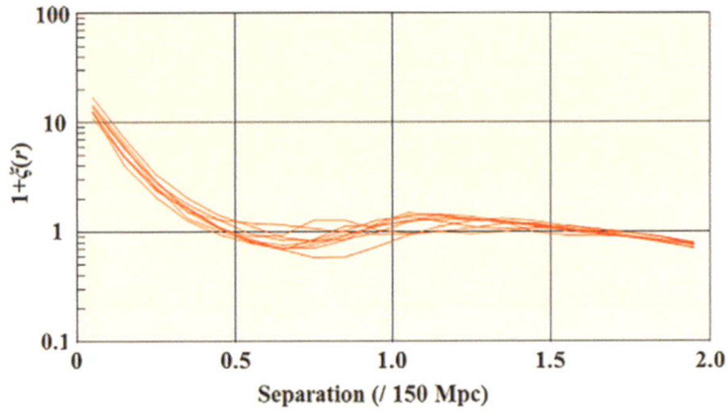

Figure 24.
Two-particle galaxy correlation function.

14. The lithium problem

We noticed as part of our original development, [3], that certain reactions were not included in the standard BBN code and that when one in particular, $Li^7 + p -> He^4 + He^4 + \gamma$, was included, the predicted abundance of lithium was lower by a factor close to three. We have since come to understand, however, that missing reactions are not the most important part of the issue. We now know that the lithium problem is another issue that arises in large part because of the incorrect standard model assumption of a uniform matter distribution at the time of nucleosynthesis. We have just shown that to understand the existence of cosmic structures and the CMB, the initial densities of protons and neutrons must have varied from place to place on all length scales. The result is that the relative abundances of elements produced at any one place will vary with the location because of variations in the initial proton and neutron densities.

The second part of the puzzle is that the observations of the element abundances are not all taken from a single location. For example, deuterium abundances have been determined from line-of-sight observations out to distances of 100 parsecs with a result of $D/H = 2.5 \times 10^{-5}$ but observations at much larger distances yield a value as large as 3×10^{-4}. (We note that one could with equal justification refer to the issue as the deuterium problem.) In **Figure 25**, we show the calculated abundances as a function of the initial baryon density, [4]. The horizontal lines are an indication of the accepted observed abundances.

(The curves in this figure differ somewhat from those shown in [4]. The reason for this are first, that a value of $H_0 = 67.3$ used in [4] instead of 73 and second, that a few changes were made to the code that improves the numerical stability of the solutions for small densities. The main effect of the latter was to change the shape of the Li^7 peak.)

From this figure, we see that The He^4 abundance is independent of the initial density over the range considered here and consequently is only dependent on the proton/neutron density ratio. The observed He^4 mass ratio is 0.245 and with $H_0 = 73$ we find that $(p/(p + n)) = 0.875$.

Considering now the deuterion and lithium abundances, we see that an initial baryon density of about five is necessary to match the accepted deuterium abundance but that a density of about 1 is necessary to account for the Li^7 abundance so the two constraints are incompatible. We noted that measurements of the deuterium abundance

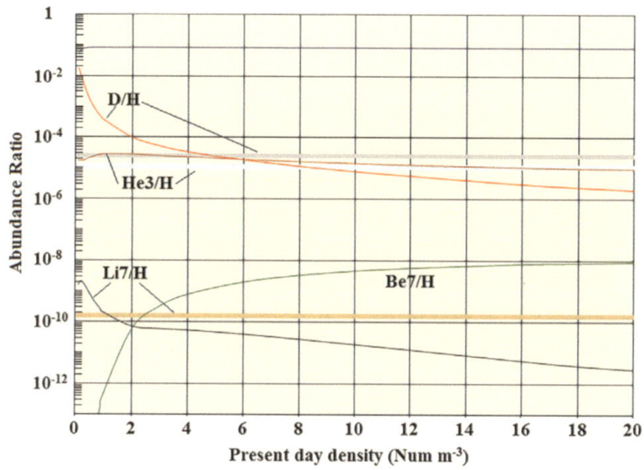

Figure 25.
Nucleosynthesis abundances as a function of the initial baryon density. $H_o = 73$.

in more distant regions where the initial density was lower yield a value considerably larger than the standard value and, in fact, the value 3×10^{-4} was obtained from sources far from any galaxy (see references in [4].) Similarly, the Li^7 abundance was determined from observations of low metal stars located far from the main disk of the Milky Way so a low initial density is again perfectly understandable. We also show results for He^3 but since that element is very active in stars, it is questionable as to whether or not the reported abundance has any relationship to its primordial value.

We find that the new model with density variations can readily account for the observed abundances although at this point, we are not able to accurately predict the detailed initial density distribution of the Milky Way to be. The lithium problem is not a problem at all. *It is only a problem if one believes that the initial baryon density was the same everywhere at the time of nucleosynthesis.*

Aside from not yet being able to predict with any accuracy the initial density profile of the matter distribution of structures to be, an additional significant issue is that many of the needed reaction cross sections are not known with great accuracy. Most of the needed low energy cross sections were measured many decades ago and soon after, experimentalists moved on to higher energies. Our simulations show that choosing one data set over another can result in significant changes to the abundance curves. Our runs, however, do show that the general pattern shown in the figure remains the same. There have not been any recent efforts made to improve the situation and additional experiments are needed to fill in the gaps and resolve differences between existing cross section determinations.

15. Rotation

A long-standing although generally neglected problem in astrophysics is the source of rotation of cosmic structures. All galaxies and some galaxy clusters rotate and even though galaxies are independent entities, their rotations are not random as is evidenced by the Tully-Fisher relationship. We spoke at the beginning about the high degree of organization that exists in the universe and the rotation of galaxies is an example of that.

The standard model idea is that the infalling material during the accretion phase of galaxy development somehow managed to start the rotation. That, however, is a highly unlikely scenario because random events do not result in organized development and they certainly could not explain the correlation between the rotation rates of different galaxies.

If we assume that the rotation was initiated at the time of nucleosynthesis and that the rotation was a matter of material undergoing normal orbital motion, the rotation would have slowed as the universe expanded. But given the present-day rotation rates and working backward, we find that the initial velocities would have been orders of magnitude greater than the speed of light which is impossible. Another problem with the accretion idea is that, as we have seen, the outer regions of developing structures had no knowledge of the center until very late in their evolution which means that organizing the material into normal orbital motion could not have begun at all until sometime after the structure had evolved beyond the causality limit. By that time, however, no process could have then started the entire structure rotating.

All this indicates that the rotation is primordial. Since the rotation cannot be explained in terms of material motion, the only other possibility is that the vacuum imprint was rotating and the matter and antimatter created by the vacuum were simply carried along by the rotating vacuum. This brings us to the reality of dark matter.

16. Dark matter

Dark matter was originally proposed to explain the motion of galaxies in galaxy clusters and later to explain the velocity distribution of the stars in spiral galaxies. Since then, it has become something of a catch-all to fix up the calculations of various cosmic phenomena that seem to defy a simple explanation.

One of the manifestations of dark matter concerns the velocity distribution of stars in spiral galaxies and the gas making up HI rings, [3, 6]. The spiral galaxy problem is illustrated by the curves in **Figure 26**.

The problem is that the observed distribution, curve B does not match the expected distribution, curve A. The generally accepted solution for this problem is to suppose there is a halo of dark *matter* surrounding the galaxy that provides the gravitation needed to match the observed velocity distribution. There are several problems with this proposal, however, not the least of which is the fact that a dark matter halo should act as a halo of stars with the lights turned off so its velocity

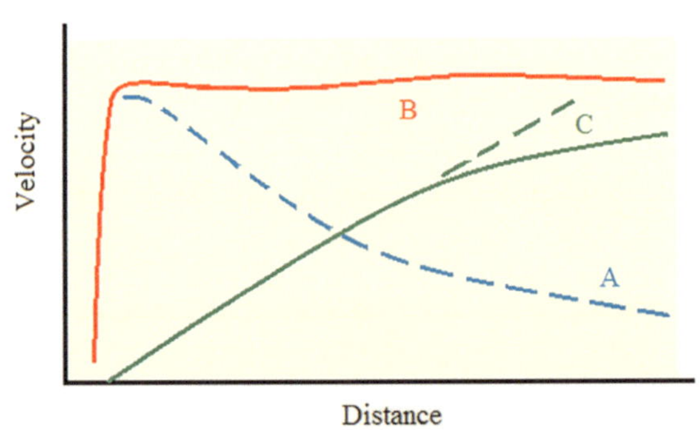

Figure 26.
Typical spiral galactic velocity distribution.

distribution should match curve A instead of B. To get a hint about the solution, we subtract the two curves to obtain curve C and notice that it approximates the velocity profile of a rotating rigid body. What this indicates is that the observed velocity distribution of the stars can be understood in terms of some normal orbital motion riding in a rigidly rotating background.

Turning to Einstein's equations, it is reasonable to model spiral galaxies using a stationary axisymmetric metric,

$$\begin{aligned} ds^2 &= -A(cdt)^2 + B(d\phi - \omega dt)^2 + Cdr^2 + Dd\psi^2 \\ &= -\left(A - \frac{B\omega^2}{c^2}\right)(cdt)^2 \\ &-2\frac{B\omega}{c}d\psi(cdt) + Bd\phi^2 + Cdr^2 + Dd\psi^2 \end{aligned} \tag{28}$$

with an energy-momentum tensor of the form

$$\mathbf{T}^{\mu\nu} = \left(\rho_{vac}c^2 + p_{vac}\right)\frac{u^\mu u^\nu}{c^2} + p_{vac}\mathbf{g}^{\mu\nu} + \rho_m c^2 \frac{v^\mu v^\nu}{c^2}. \tag{29}$$

A small volume of the vacuum will respond to the total gravitation field in the same way as a material particle which means that we can analyze its motion using the usual geodesic equations. The first two of these are satisfied identically as a consequence of our assumption of a stationary metric. The remaining two have the solution

$$\dot{\varphi}_{vac}(r,\psi) = \omega(r,\psi) \tag{30}$$

where $\omega(r,\psi)$ represents the rotation of the galaxy. We find that the vacuum must rotate and, as shown in [3], it must do so with zero angular momentum. Because of that fact, the rotation is not subject to the slowing induced by the expansion as would be normal orbital motion.

We now consider the stars whose velocity must also satisfy the geodetic equations. We separate their motion into a component that is at rest in the vacuum, and hence without angular momentum, and a residual with normal orbital motion, $\dot{\phi}_m = \dot{\phi}_{m,r} + \overline{\omega}$. Solving the geodetic equations gives

$$\dot{\phi}_{m,r}(l,z) = \frac{v}{l} - \overline{\omega}(l,z) \tag{31}$$

which exhibits the behavior shown in **Figure 26**. On the left is the normal orbital rotation, curve A, the first term on the right is the constant velocity component, curve B, and the second is the rigid body vacuum rotation, curve C. At the outer edge of the galaxy, the constant stellar velocity $v = \overline{\omega}(l_G, 0)$ equals the rotation rate so $\dot{\phi}_{m,r}(l_G, 0) = 0$ which means that the outermost stars are at rest in a rotating vacuum.

This result dovetails quite nicely with the rotation model discussed in the previous section. In that case, the vacuum rotation carried along the material and, in this case, the material appears to drag along the vacuum but the result is the same. The vacuum rotation is a consequence of the Planck era imprint but it is likely that the residual orbital motion did not begin until after the ZVPs were reached and the galaxies began to collapse.

Another problem mentioned in [1] is known as the "angular momentum catastrophe." The idea behind this thoery is that the development of disk galaxies, and in particular their sizes, can be understood in terms of their angular momenta [9]. The problem is that the theory does not work. The predicted sizes are an order of magnitude too small. There are a number of assumptions made that we do not accept such as the idea that their rotations are a consequence of tidal torques between neighboring galaxies instead of being primordial. The reason for that is that gravitation had no effect on the evolution of galaxies until very shortly before they began their initial collapse and certainly could not have exerted enough torque on neighboring galaxies to start them rotating within the short time interval prior to their ZVP. All galaxies rotate and it is highly unlikely that this mechanism could account for their rotations considering the variations in their separations and orientations. The reason for putting these comments in this section, however, concerns the assumption that the angular momentum of the stars increases with distance from the center of the galaxy. In this new model the exact opposite is true. The outer stars in a disk are carried along by the rotation of the vacuum and do not have any angular momentum at all.

In the references, we discuss other cases in which vacuum energy is shown to account for phenomena attributed to dark matter. Summing up, *when the vacuum is included in Einstein's equations, the dark matter problem disappears*.

17. Hubble tension

During the past two decades or so, a lot of effort has gone into the problem of pinning down the value of the Hubble constant. The various methods can be grouped according to the redshift range involved. First are those based on observations in the range from about 0.01 to an upper limit of 6 or so. The significance of the lower limit is that it is the smallest redshift for which the recession velocities of galaxies have a reasonable probability of being significantly larger than their peculiar velocities. The upper limit just represents the time of galaxy formation. The Hubble constant determined by methods covering this range is settling on a value of $H_0 \approx 73$, [8].

In **Figure 27**, we show new model luminosity distance modulus predictions for 3 values of the Hubble constant. As we see, there is only a very slight dependence on the

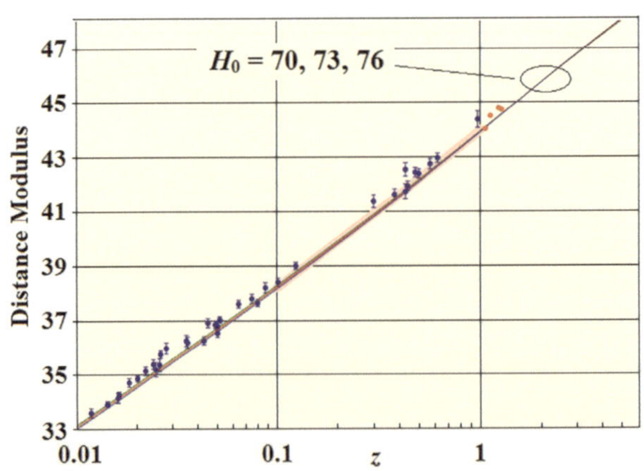

Figure 27.
Luminosity distance predictions of the new model. Refer to [3] for a detailed description of the data points.

Hubble constant. It is important to appreciate that the new model has no other adjustable parameters and that varying the Hubble constant changes the result hardly at all so a failure to fit the data would be a real problem for the model. It appears that the model predictions lie a little below the data points but the figure shows only a very limited number of the total number of data points now in existence. When the new points are included, the new model fit is extremely good.

We now compare with the ΛCDM predictions which are shown in **Figure 28**. We see that the ΛCDM model shows a greater dependence on the value of the Hubble constant than does the new model. The predictions of the two models are, however, in no way of equal significance. As already noted, the new model prediction cannot be adjusted. The ΛCDM model results, on the other hand, aren't predictions at all but instead are curve fits with three parameters; Ω_m, Ω_Λ, and the Hubble constant.

The Riess SHOES group finds a value of $H_0 = 73.3$ for the FRW parameters just indicated. In **Figure 29** we see that with $H_0 = 73$, the results are essentially identical for redshifts somewhat smaller than two. Since the value $H_0 = 73$ gives the best fit of the data in the ΛCDM case, it validates the prediction of the new model.

The second grouping of methods covers the redshift range from 0.01 down to some lower cutoff which we will take to be 0.001. The overriding factor in this redshift range is that the galactic pecular velocities are as large or larger than the recession velocities. The best-known results covering this redshift range are those of the Tip of the Red Giant Branch (TRGB) group. They claim that their data results in a Hubble constant of $H_0 = 69.8$ which comprises the second largest contributor to the Hubble tension. There are, however, problems with that result. First, the sample set used contained only 18 galaxies, and second, the results are very sensitive to the statistical model used to separate the recession velocity of each galaxy from its observed velocity.

In [8], we did a fit using *all* the data points in the TRGB database that had observed velocities in the redshift range we are considering. The result was a total of 291 galaxies. To determine the Hubble constant, the recession velocity of each galaxy has to be separated from the observed velocity and at present, the only way this can be done is by using a statistical model which is based on the mass distribution of the local universe. Tools that implement this process are available online in the Extragalactic Distance Database, https://edd.ifa.hawaii.edu/ which also contains the TRGB database.

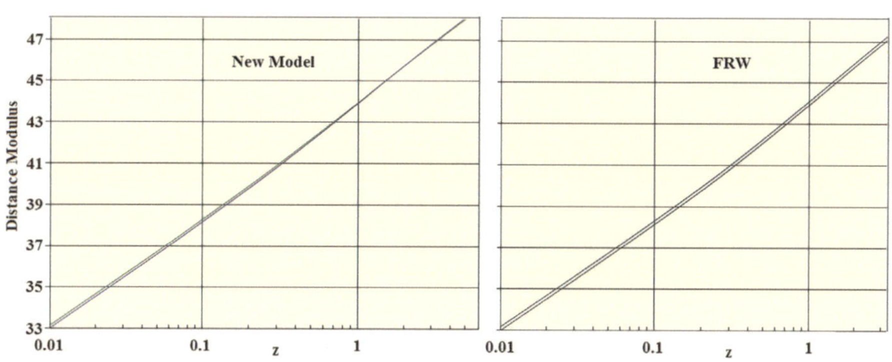

Figure 28.
Model and FRW predictions for $H_o = 70$ and $H_o = 76$. The FRW parameters are $\Omega_m = 0.3$ and $\Omega_\Lambda = 0.7$.

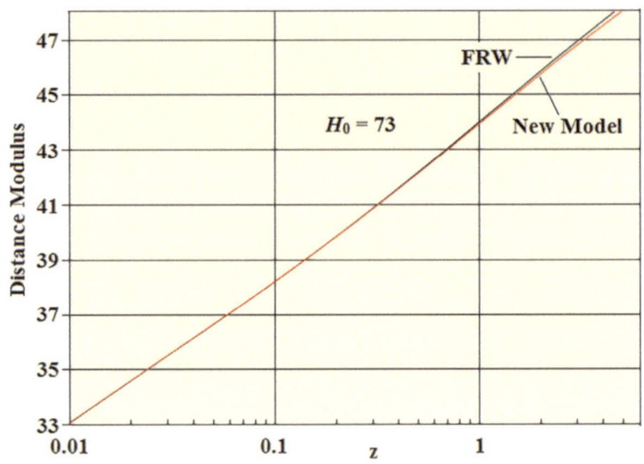

Figure 29.
New model (red) and FRW (black) predictions for $H_o = 73$.

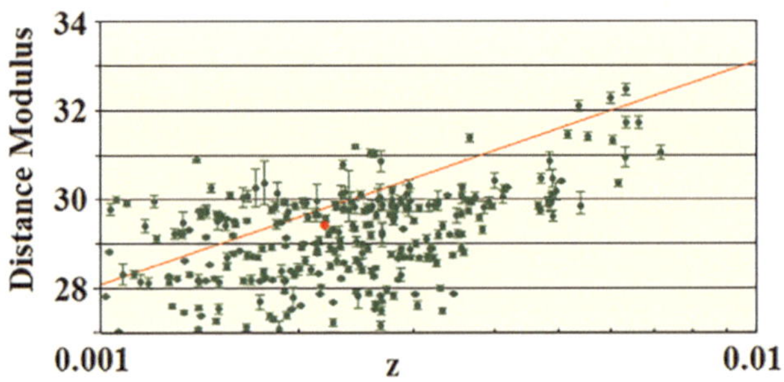

Figure 30.
TRGB raw data points.

In **Figure 30**, we show the raw data points and in **Figure 31**, the same set corrected for the peculiar velocities using the probabilistic model. The red line is the new model prediction. The difference between the two distributions makes it obvious that the peculiar velocities are large so any prediction made is highly dependent on the accuracy of the correction.

If the model were completely accurate, all the adjusted data points would lie along a single line. Since they do not, there is obviously a problem. We note that the new

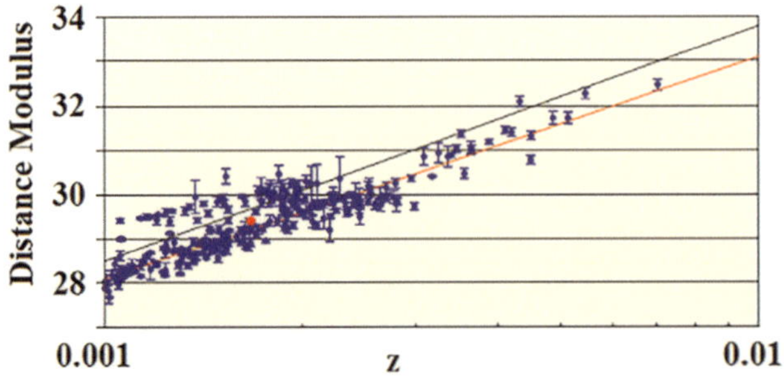

Figure 31.
TRGB data points adjusted for peculiar velocities.

model prediction gives a good fit to what appears to be the correctly adjusted galaxies. In **Figure 32**, we show normalized χ^2 results for two cases. In the upper panel, we show the result for the full data set of **Figure 30**. This indicates a Hubble constant in the range 66–67 which is not a lot different from the group's published value. In **Figure 31**, we have included a black line that roughly separates the spurious points from the primary set. In the lower panel of **Figure 31**, we show the result when the data points above the black line are removed. This time the result is a Hubble constant close to 73 in agreement with the supernovae determinations so clearly, it is the peculiar correction model that is being tested rather than the distance determinations and with perfect corrections, there would not be any tension.

The third redshift grouping, commonly referred to as cosmological, groups together those methods covering redshifts greater than that of galaxy formation. The BAO model of the CMB spectrum falls under this heading and it is the only such model making a Hubble constant prediction. The consensus value is $H_0 \approx 67$ which constitutes the biggest portion of the Hubble tension. We have already shown, however, that the BAO model is wrong so this Hubble constant value can be dismissed.

There are other approaches such as the Megamaser and $L - \sigma$ methods which are discussed in [8]. The results from those two methods are consistent with a value of 73. Still other methods are discussed in [1].

There is one further approach that has the potential for becoming a viable method for determining the Hubble constant. The idea is to measure the rate of recession of bodies within the solar system. At present, there is considerable interest in Moon-Earth distance because 70 years of accurate distance measurements already exist. Calculating the Hubble flow contribution to the total recession rate is very simple and the result shows that the rate is well within a range that can be measured. The difficulty comes from the fact that all the other contributors to the Moon's recession, in particular the tidal effects, must be determined with an combined error of no more than 1 or 2 percent before a Hubble determination can be made.

Planetary recession offers other opportunities. Instead of the 3.8 cm per year recession of the Moon, the rates would be 10 meters per year or more depending on

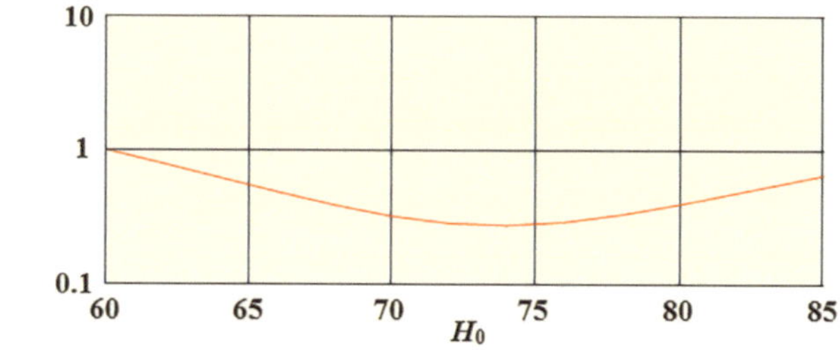

Figure 32.
*χ^2 for the data points in **Figure 31**. The upper panel is the result with all the data points included and the lower panel with the points above the black line removed.*

the planet. Measuring the distance becomes more difficult but, on the plus side, the planets are not as subject to the large tidal effects that make the Moon determination difficult.

Summing up, of the two principal contributors to the tension, one is just wrong and the other did not include the entire data set or properly make allowances for errors in the peculiar velocity prediction model. When the spurious values are removed, the results yield the same value as that of the larger z studies. The conclusion is that *the tension has been eliminated and the correct value is very close to* $H_0 = 73$.

18. Tying things together

We have shown that accretion initiated by small fluctuations in an otherwise uniform distribution of ordinary matter is impossible. We also showed that gravitation was ineffective until shortly before t_G so the idea that accretion is the primary source of cosmic structures is wrong. Another insurmountable problem with accretion is that no process that involves communication could account for structures as large as superclusters. The conclusion we reached is that the existence of structures is a consequence of an imprint established during the initial Planck era inflation. Since the imprint regulated the material manifestation of the structures at the time of nucleosynthesis, **Figure 19** is not just a map of the CMB anisotropy but is also a photograph of the vacuum as it existed at the end of the inflation.

This, however, leaves us with the question of what organized the imprint. It could not have been some random process so there must have been some rule that regulated its development. We know that causality played no part in the development and that whatever the rule was, it must have been extremely simple. In terms relative to the size of the expanding universe, the imprint structures ranged from the size of a star up to the size of superclusters with the latter being nearly as large as the universe at the end of the inflation and the beginning of the transition. The energy densities of these structures, on the other hand, were on the order of $10^{-4} - 10^{-5}$ relative to the average so they were large but faint.

In **Figure 33**, we show a plot of the count of cosmic structures versus their size, [3].

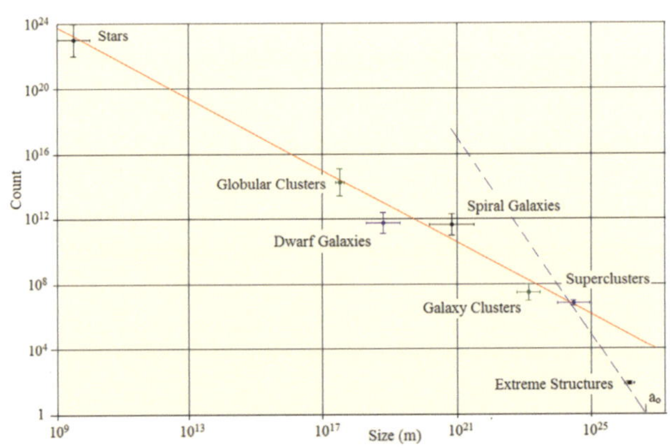

Figure 33.
Count of structures vs. size.

What is remarkable is that aside from the extreme structures, all cosmic structures with their vast differences in size and numbers lie on a power-law curve and this holds all the way down to the stars. The extreme structures (size $> 45°$) fall below this line but this is simply a consequence of the finite size of the universe. Another point to notice is that the structures have distinct sizes with no overlap. If accretion was the process by which the structures were formed, one would expect a continuum of sizes instead. A fit to the curve has the following result

$$C(s) = 5.7 \times 10^6 \left(\frac{s_{sc}}{s}\right)^{1.1} \tag{32}$$

where we have scaled by the average size of a supercluster. The dashed line shows the count of structures of a given size that would fill the universe and its formula is

$$C_{filled}(s) = (a_0/s)^3. \tag{33}$$

We see that superclusters fall on both lines so, in an order of magnitude sense, they fill all space.

The significant factor in the second formula is the power of three. The idea of a fractal dimension extends this concept to situations in which the power can have any value, not just an integer but this is exactly the form of (Eq. (32)) so we find that the structure imprint had a fractal dimension of about 1.1. It is generally true that the fractal dimension of any system is larger than the geometric dimension of that same structure so it follows that the cosmic structure must be 1-dimensional or in other words, that the *structure of the universe must be one of filaments* which gives us the cosmic web.

A defining property of fractal structures is self-similarity and this suggests that self-similarity was in some way the defining principle that regulated the formation of the imprint.

19. Conclusions

We have summarized the results contained in a series of papers published during the past 3 years. A new model of cosmology based on a universe with time-varying curvature is presented that makes a large number of predictions that agree with observation. This model includes a new model of nucleosynthesis that accounts for the existence of all cosmic structures and the CMB, the origin of galactic supermassive black holes, and the stability of galaxy clusters. It also explains the existence of very large galaxies at a time at which the standard model says they could not yet exist.

We have shown that the problems with the standard model boil down to three incorrect assumptions and one omission; the first incorrect assumption is that the curvature of spacetime is constant, the second is that the initial distribution of matter was uniform, and the third is that structure formation was a results of accretion. When those assumptions are abandoned and vacuum energy is included in the energy-momentum tensor, everything falls into place.

Here at the end, we are back at the beginning with two problems to solve. The first is to understand the evolution of the Planck era inflation. We have no clear idea of how the fractal imprint came into existence but we have demonstrated that it was responsible for the universe we observe. The second problem is to understand the breaking of the baryon conservation law that results in the creation of the very small percentage of extra neutrons at the time of nucleosynthesis.

Author details

John C. Botke
Ronin Institute, Montclair, USA

*Address all correspondence to: jcbotke@ronininstitute.org

References

[1] Perivolaropoulos L, Skara F.. Challenges for ΛCDM: An update. arXiv: 2105.05208v3. 6 Apr 2022

[2] Abdalla E, et al. Cosmology intertwiined: A review of the particle physics, astrophysics, and cosmology associated with the cosmological tensions and anomalies. arXiv: 2203.06142vf1. 2022

[3] Botke JC. A different cosmology - thoughts from outside the box. Journal of High Energy Physics Gravitation and Cosmology. July 2020;**6**(3):473-566

[4] Botke JC. The origin of cosmic structures part 4 – nucleosynthesis. Journal of High Energy Physics, Gravitation and Cosmology. 2022;**8**(3): 768-799

[5] Botke JC. The origin of cosmic structures part 1 - stars to superclusters. Journal of High Energy Physics, Gravitation and Cosmology. 2021;**7**(4): 1373-1409

[6] Botke JC. The origin of cosmic structures Part 2 - HI rings. Journal of High Energy Physics, Gravitation and Cosmology. 2021;**7**(4):1410-1424

[7] Botke JC. The origin of cosmic structures Part 3 - Supermassive black holes and galaxy cluster evolution. Journal of High Energy Physics, Gravitation and Cosmology. 2022;**8**(2): 345-371

[8] Botke JC. The origin of cosmic structures Part 5 - Resolution of the hubble tension problem. Journal of High Energy Physics, Gravitation and Cosmology. 2023;**9**(1):60-82

[9] van den Bosch F, C, Burkert A, Swaters RA. The angular momentum content of Dwarf galaxies: New challenges for the theory of galaxy formation. Monthly Notices of the Royal Astronomical Society. 2001;**326**(3): 1205-1215

Chapter 5

Rotating Cosmological Models, or How Matter and Light Can Escape from a Rotating Black Hole

Herwig Dejonghe

Abstract

We show how to construct models for rotating black holes with comoving matter inside, and how they canc be interpreted as models for rotating universes. The metric along the polar axis of such a black hole is fundamentally different from the metric elsewhere inside the black hole. We show that this creates the possibility that matter or light can escape through the poles. As a corollary, we show how the Lemaitre-Tolman metric can be obtained from the Schwarzschild metric.

Keywords: Einstein equations, exact solution, cosmology, black holes, rotation

1. Introduction

The motive for this paper is the statement in Dejonghe [1] that matter and light may be able to escape a rotating black hole through the poles. The rationale behind that statement was that in a nonrotating nonempty black hole (i.e., a black hole containing a distributed mass density, not only containing a central point mass), the accompanying metric allows this phenomenon under certain conditions. These cannot realistically be met globally in a spherical symmetry, however, which was the framework [1] was written in. However, there is some similarity between the Schwarzschild metric and the Kerr metric, the latter taken along the polar axis. Since the polar axis is one-dimensional, the conditions mentioned below could be met in a Kerr metric along the polar axis. This statement was but a conjecture in Ref. [1]. The trigger of this paper was to prove it.

Sections 2 and 3 set the stage and do not contain new material. The core of the paper is to be found in sections 4 and 5. Section 6 recapitulates [1], but here from a different perspective. Section 7 contains the framework for the construction of models for a nonempty rotating black hole, that can also be seen as cosmological models for rotating universes in the presence of a cosmological constant. Section 8 reconnects with Ref. [1] and proves that matter and light can escape from a rotating nonempty black hole through the poles.

2. The Kerr–de Sitter metric

The Kerr–de Sitter solution (hereafter KdS) of Einstein's equations represents a rotating black hole with mass $\mathcal{M}$ in a world with a cosmological constant Λ. It is the generalization of the Kerr geometry, which has two parameters: the geometrical mass of the black hole $M = G\mathcal{M}/c^2$ and a rotation parameter $a = L_z/(\mathcal{M}c)$, with L_z an angular momentum. Both parameters have the dimension of a length. We follow here largely Hackmann et al. [2] and denote as usual

$$\rho^2(r, \vartheta) = r^2 + a^2 \cos^2\vartheta = r^2 + a^2 - a^2 \sin^2\vartheta, \tag{1}$$

$$\Delta_r(r) = r^2 - 2Mr + a^2 - \frac{\Lambda}{3} r^2 (r^2 + a^2) \quad \text{and} \quad \Delta_\vartheta(\vartheta) = 1 + \frac{\Lambda}{3} a^2 \cos^2\vartheta. \tag{2}$$

The metric reads

$$ds^2 = g_{00} dt^2 + g_{11} dr^2 + g_{22} d\vartheta^2 + g_{33} d\varphi^2 + 2g_{03} dt d\varphi, \tag{3}$$

with

$$g_{00} = \frac{1}{\rho^2} (\Delta_r - a^2 \sin^2\vartheta \Delta_\vartheta), \quad g_{11} = -\frac{\rho^2}{\Delta_r}, \quad g_{22} = -\frac{\rho^2}{\Delta_\vartheta}, \tag{4}$$

$$g_{33} = \frac{\sin^2\vartheta}{\rho^2} \left[\Delta_r a^2 \sin^2\vartheta - \Delta_\vartheta (r^2 + a^2)^2\right] \quad \text{and} \quad g_{03} = \frac{a \sin^2\vartheta}{\rho^2} \left[\Delta_\vartheta (r^2 + a^2) - \Delta_r\right]. \tag{5}$$

The coordinate 'time' t has the dimension of a length. We note that the parameters M and Λ are tucked away in the mutually independent functions $\Delta_r(r)$ and $\Delta_\vartheta(\vartheta)$.

In the sequel, we will also use the following shorthand notation, inspired on (1):

$$\rho'^2(r, \vartheta) = r^2 + a^2 + a^2 \sin^2\vartheta. \tag{6}$$

3. The orbits

3.1 The general case $a \neq 0$

It is essential for this paper that the Hamilton–Jacobi equation for geodesic motion, which reads (see also Chandrasekhar [3])

$$2\partial_\tau S = g^{\mu\nu} \partial_{x^\mu} S \partial_{x^\nu} S \tag{7}$$

with ∂_{x^μ} the partial derivative with respect to variable x^μ, τ the proper time and S Hamilton's principal function, is separable in de KdS spacetime. Its solution is

$$S = \frac{c^2}{2}\delta\tau - cEt + cL_z\varphi + \varepsilon_r c \int \frac{\sqrt{R(r')}}{\Delta_r(r')} dr' + \varepsilon_\vartheta c \int \frac{\sqrt{\Theta(\vartheta')}}{\Delta_\vartheta(\vartheta')} d\vartheta'. \tag{8}$$

In this equation $\delta = 0$ for light and $\delta = 1$ for material particles, in which case E is the energy normalized to the rest energy $m_0 c^2$ (and thus dimensionless) and L_z is the angular momentum normalized with $m_0 c$ (and thus with the dimension of a length), while, also here, t has the dimension of a length. The integrals stand for primitives of the integrands. The sign constants ε_r and ε_ϑ equal ± 1, independent of each other. Their signs equal the signs of the derivatives of r and ϑ (as functions of τ). The functions $R(r)$ and $\Theta(\vartheta)$ are given by

$$R(r) = \left[E\left(r^2 + a^2\right) - aL_z\right]^2 - \left(\delta r^2 + a^2 K\right)\Delta_r(r) \tag{9}$$

and

$$\Theta(\vartheta) = -\frac{1}{\sin^2(\vartheta)}\left[aE\sin^2\vartheta - L_z\right]^2 + a^2\left(K - \delta\cos^2\vartheta\right)\Delta_\vartheta(\vartheta). \tag{10}$$

In both these equations appears, apart from E and L_z, the integration constant a^2K which follows from the separability. Note that our definition of K differs from the usual one by a factor a^2, but was chosen here in order that K has the same dimension as E, and thus dimensionless in our units[1] ($a^2 K m_0 c^2$ has the dimension of a square of an action).

In this paper we will only be dealing with orbits that have $L_z = 0$. Since φ does not enter the equations, what we call here an orbit is in fact a filled ring centered on the polar axis. We denote an orbit in the ring with $(\bar{t}(\tau), \bar{r}(\tau), \bar{\vartheta}(\tau), \bar{\varphi}(\tau))$, here with omitted reference to the integration constants E and K.

Our formulae will be explicit for infalling orbits (hence $\varepsilon_r = -1$). For outgoing orbits (for as far as this makes sense) $\sqrt{R}$ has to be changed to $-\sqrt{R}$ (hence $\varepsilon_r = +1$). The quantity $\sqrt{\Theta}$ can have both signs, depending, for infalling orbits, on whether the orbit is 'above' ($\varepsilon_\vartheta = +1$, $0 \le \vartheta \le \pi/2$) or 'below' ($\varepsilon_\vartheta = -1$, $\pi/2 \le \vartheta \le \pi$) the equatorial plane (and vice versa for outgoing orbits). Here we choose $\varepsilon_\vartheta = +1$.

We will need the equations for these orbits in differential form as well as in quadrature form. The differential equations are (for completeness the terms in $L_z \neq 0$ are also shown between brackets):

$$\begin{aligned}
\rho^2\dot{r} &= \rho^2 r_\tau = -c\sqrt{R} \\
\rho^2\dot{\vartheta} &= \rho^2\vartheta_\tau = c\sqrt{\Theta} \\
\rho^2\dot{\varphi} &= \rho^2\varphi_\tau = cEa\left[\frac{r^2 + a^2}{\Delta_r} - \frac{1}{\Delta_\vartheta}\right]\left(+\frac{L_z}{\sin^2\vartheta}\left[\frac{1}{\Delta_\vartheta} - \frac{a^2\sin^2\vartheta}{\Delta_r}\right]\right) \\
\rho^2\dot{t} &= \rho^2 t_\tau = cE\left[\frac{\left(r^2 + a^2\right)^2}{\Delta_r} - \frac{a^2\sin^2\vartheta}{\Delta_\vartheta}\right]\left(+aL_z\left[\frac{1}{\Delta_\vartheta} - \frac{r^2 + a^2}{\Delta_r}\right]\right),
\end{aligned} \tag{11}$$

with the dot differentiation with respect to τ. The functions $R(r)$ and $\Theta(\vartheta)$ now ($L_z = 0$) read:

$$R(r) = E^2\left(r^2 + a^2\right)^2 - \left(r^2 + a^2 K\right)\Delta_r(r) \ge 0 \tag{12}$$

and

[1] When $a = 0$, the constant K disappears anyway and can be replaced by ϑ.

$$\Theta(\vartheta) = -a^2E^2\sin^2\vartheta + a^2\left(K - \cos^2\vartheta\right)\Delta_\vartheta(\vartheta) \geq 0. \tag{13}$$

The quadratures are obtained by differentiation of S with respect to the constants of the integration. For infalling orbits above the equatorial plane, Hamilton's principal function can be written as

$$S = \frac{c^2}{2}\delta(\tau - \tau_0) - cE(\bar{t} - \bar{t}_0) + cL_z(\overline{\varphi} - \overline{\varphi}_0) + c\int_{\bar{r}}^{\bar{r}_0}\frac{\sqrt{R(r)}}{\Delta_r(r)}dr + c\int_{\bar{\vartheta}_0}^{\bar{\vartheta}}\frac{\sqrt{\Theta(\vartheta)}}{\Delta_\vartheta(\vartheta)}d\vartheta, \tag{14}$$

with $\bar{t}_0 = \bar{t}(\tau_0), \bar{r}_0 = \bar{r}(\tau_0), \overline{\vartheta}_0 = \overline{\vartheta}(\tau_0)$ and $\overline{\varphi}_0 = \overline{\varphi}(\tau_0)$ the coordinates of any point along an orbit. Upon partial differentiation of S with respect to K, δ, E and L_z we obtain respectively

$$a^2\int_{\bar{r}}^{\bar{r}_0}\frac{dr}{\sqrt{R(r)}} = a^2\int_{\bar{\vartheta}_0}^{\bar{\vartheta}}\frac{d\vartheta}{\sqrt{\Theta(\vartheta)}} \tag{15}$$

$$c(\tau - \tau_0) = \int_{\bar{r}}^{\bar{r}_0}\frac{r^2dr}{\sqrt{R(r)}} + a^2\int_{\bar{\vartheta}_0}^{\bar{\vartheta}}\frac{\cos^2\vartheta d\vartheta}{\sqrt{\Theta(\vartheta)}} \tag{16}$$

$$\bar{t} - \bar{t}_0 = E\int_{\bar{r}}^{\bar{r}_0}\frac{\left(r^2 + a^2\right)^2dr}{\sqrt{R(r)}\Delta_r(r)} - a^2E\int_{\bar{\vartheta}_0}^{\bar{\vartheta}}\frac{\sin^2\vartheta d\vartheta}{\sqrt{\Theta(\vartheta)}\Delta_\vartheta(\vartheta)} \tag{17}$$

$$\overline{\varphi} - \overline{\varphi}_0 = aE\int_{\bar{r}}^{\bar{r}_0}\frac{(r^2 + a^2)dr}{\sqrt{R(r)}\Delta_r(r)} - aE\int_{\bar{\vartheta}_0}^{\bar{\vartheta}}\frac{1}{\sqrt{\Theta(\vartheta)}\Delta_\vartheta(\vartheta)}d\vartheta \tag{18}$$

The first equation suffices to determine $\overline{\vartheta}(r)$, a function depending on choices for E, K and (r_0, ϑ_0) (apart from also depending on a, M and Λ, of course, but we will omit explicit reference to these latter dependencies). The second will in addition yield, be it implicitly, $\bar{r}(\tau, E, K)$ and $\overline{\vartheta}(\tau, E, K)$ after choosing τ_0.

The other two have right hand sides that tend to $+\infty$ at the horizons $\Delta_r = 0$ or $\Delta_\vartheta = 0$, making t and φ unsuitable as coordinates that describe the orbits there. In our final results, however, $\bar{t}$ and $\overline{\varphi}$ will have disappeared altogether. Eqs. (14)–(18) can easily be modified to accommodate infalling orbits below the equatorial plane and outgoing orbits above and below the equatorial plane.

In the sequel, we will frequently need the partial derivatives with respect to E and K of all functions $\bar{t}(\tau), \bar{r}(\tau), \overline{\vartheta}(\tau)$ and $\overline{\varphi}(\tau)$. We will denote these derivatives by the coordinate with added subscript (but no bar, for simplicity). We will not need explicit expressions for r_E, r_K, ϑ_E and ϑ_K. It can be shown that r_E and r_K change sign with ε_r, and likewise ϑ_E and ϑ_K with ε_ϑ. The expressions for φ_E and φ_K are obtained from (17):

$$\varphi_E = -\frac{aE\left(\bar{r}^2 + a^2\right)}{\sqrt{R}\Delta_r}r_E - \frac{aE}{\sqrt{\Theta}\Delta_\vartheta}\vartheta_E + w_E\left(\bar{r}, \overline{\vartheta}\right) \tag{19}$$

and

$$\varphi_K = -\frac{aE\left(\bar{r}^2 + a^2\right)}{\sqrt{R}\Delta_r}r_K - \frac{aE}{\sqrt{\Theta}\Delta_\vartheta}\vartheta_K + w_K\left(\bar{r}, \overline{\vartheta}\right) \tag{20}$$

with

$$w_E(\bar{r}) = -a\int_{\bar{r}}^{\bar{r}_0} \frac{(r^2+a^2)(r^2+a^2K)}{R^{3/2}(r)}dr - a^3\int_{\bar{\vartheta}_0}^{\bar{\vartheta}} \frac{K - \cos^2(\vartheta)}{\Theta^{3/2}(\vartheta)}d\vartheta$$
$$w_K(\bar{r}) = \frac{a^3E}{2}\int_{\bar{r}}^{\bar{r}_0} \frac{r^2+a^2}{R^{3/2}(r)}dr + \frac{a^3E}{2}\int_{\bar{\vartheta}_0}^{\bar{\vartheta}} \frac{1}{\Theta^{3/2}(\vartheta)}d\vartheta. \tag{21}$$

The expressions for t_E and t_K are calculated with the aid of (14):

$$Et_E = -\frac{\sqrt{\bar{R}}}{\bar{\Delta}_r} r_E + \frac{\sqrt{\bar{\Theta}}}{\bar{\Delta}_\vartheta}\vartheta_E \quad \text{and} \quad Et_K = -\frac{\sqrt{\bar{R}}}{\bar{\Delta}_r} r_K + \frac{\sqrt{\bar{\Theta}}}{\bar{\Delta}_\vartheta}\vartheta_K. \tag{22}$$

We note that also φ_E, φ_K and t_E, t_K are independent of the signs ε_r or ε_ϑ.

The problem of finding the analytic solutions of geodesic motion in the KdS spacetime has a long history and was fully solved in Ref. [2], with many references therein. In this paper, we will only touch on this difficult and specialist topic when we consider nonempty rotating black holes (or universes) with $\Lambda = 0$, this for the purpose of illustrating the procedure of actually constructing a model for a nonempty rotating black hole or universe. We will solve the equations of motion in that case in terms of elliptic functions and Carlson elliptic integrals (Appendix D).

3.2 The case $a = 0$

It follows from (13) that $\Theta = 0$, hence, with (11), ϑ and φ are constants (see also (18)). The orbits are radial $\bar{r}(\tau, E, \vartheta)$, with ϑ taking on the role of K. The latter is not undefined, since it follows from (32) that $K = \cos^2\vartheta + \sin^2\vartheta\Delta_r(r_a)/r_a^2 = \cos^2\vartheta + \sin^2\vartheta\Delta(r_a)$. In addition we find, with (2) and (12):

$$\Delta_r = r^2\left(1 - \frac{2M}{r} - \frac{\Lambda}{3}r^2\right) \equiv r^2\Delta, \qquad \Delta_\vartheta = 1,$$
$$R = r^2(E^2r^2 - \Delta_r) = r^4(E^2 - \Delta). \tag{23}$$

The equations of the motion are

$$\dot{r} = r_\tau = -c\sqrt{E^2 - \Delta} \quad \text{and} \quad \dot{t} = t_\tau = \frac{cE}{\Delta}. \tag{24}$$

Eq. (15) does not exist, since it reduces to $0 = 0$. All the orbits can reach $r = 0$. With that choice for r_0, we can recast (16) in the form

$$c(\tau - \tau_0) = -\int_0^{\bar{r}} \frac{\sqrt{r}dr}{\sqrt{2M - (1-E^2)r + \Lambda r^3/3}} = -\int_0^{\bar{r}} \frac{dr}{\sqrt{E^2 - \Delta}}, \tag{25}$$

which is the same as the defining equation for the downward branch of the function cyc in Ref. [1], Appendix 1. Of all the partial derivatives, only r_E and t_E are left in a nontrivial way. We will need (22):

$$Et_E = -\frac{\sqrt{E^2 - \bar{\Delta}}}{\bar{\Delta}}. \tag{26}$$

As for the other partial derivatives, r_K and t_K are to be replaced by $r_\vartheta = 0$ and $t_\vartheta = 0$, $\vartheta_E = 0$ and ϑ_K has to be replaced by $\vartheta_\vartheta = 1$. The functions $w_E = w_K = 0$.

3.3 The case $\vartheta = 0$

Now the integrals over ϑ are indeterminate since $\Theta = 0$, but can all be evaluated using (15). We obtain

$$c(\tau - \tau_0) = \int_{\bar{r}}^{r_0} \frac{r^2 + a^2}{\sqrt{R(r)}} dr \quad \text{and} \quad \bar{t} - \bar{t}_0 = E \int_{\bar{r}}^{r_0} \frac{(r^2 + a^2)^2}{\sqrt{R(r)}\Delta_r(r)} dr. \tag{27}$$

The equation for $\overline{\varphi}$ is, as the above integral, a proper one for $\Delta_r > 0$ and $\Delta_\vartheta > 0$:

$$\overline{\varphi} - \overline{\varphi}_0 = aE \int_{\bar{r}}^{r_0} \left(\frac{r^2 + a^2}{\Delta_r(r)} - \frac{1}{\Delta_\vartheta} \right) \frac{dr}{\sqrt{R(r)}}. \tag{28}$$

Between the brackets, we recognize an expression that is proportional to $\dot{\varphi}$, but the meaning of this last integral is a limit $\vartheta \to 0$, since φ is not defined on the polar axis. We will not need it anyway.

The orbits are radial $\bar{r}(\tau, E)$. Regarding the partial derivatives, only r_E and t_E are left. The others do not exist, since $\vartheta (= 0)$ is not a coordinate anymore and neither is $K (= 1)$. We will not need w_E nor w_K.

4. The travelator metric

4.1 The general case $a \neq 0$

We now pass from the coordinates $(x^\alpha) = (t, r, \vartheta, \varphi)$ to new coordinates $(\overline{x}^\alpha) = (\tau, E, K, \psi)$. The angle ψ is the same as φ and is additive to it, i.e. $\varphi = \overline{\varphi}(E, K, \tau) + \psi$. Such a metric is called a synchronous metric (see also Landau and Lifshitz [4]). This metric is the generalization of the Novikov metric, much as the synchronous metric of Khatsymovski [5] is the generalization of the Lemaitre metric.

Each of the old coordinates is now function of E, K, τ, and , and we adopt the transformation matrix

$$C = (C_{\alpha\beta}) = \begin{pmatrix} t_\tau & r_\tau & \vartheta_\tau & \varphi_\tau \\ t_E & r_E & \vartheta_E & \varphi_E \\ t_K & r_K & \vartheta_K & \varphi_K \\ 0 & 0 & 0 & 1 \end{pmatrix} = \left(\frac{\partial x^\beta}{\partial \overline{x}^\alpha} \right), \tag{29}$$

where t_τ, r_τ, ϑ_τ, and φ_τ are given by (11).[2] The indices α and β are simply matrix indices. The metric transforms according to $\overline{g}_{\alpha\beta} = \frac{\partial x^\rho}{\partial \overline{x}^\alpha} \frac{\partial x^\sigma}{\partial \overline{x}^\beta} g_{\rho\sigma}$, or in matrix form

$$(\overline{g}) = T(g)T^T, \tag{30}$$

[2] That the total derivative with respect to τ equals the partial derivative, follows from the fact that the orbits evolve at constant E and K.

with $\mathcal{T} = C$. The matrices (g) and $(\overline{g})$ are therefore congruent.

The determinant of C is given by a remarkably simple expression:

$$\det(C) = \frac{c}{E}(r_E\vartheta_K - r_K\vartheta_E). \tag{31}$$

It does not contain singularities connected with the event horizons $\Delta_r = 0$ or $\Delta_\vartheta = 0$. We recognize in the last factor the Jacobian of the transformation from (E, K) to $(\overline{r}, \overline{\vartheta})$, with τ as a parameter.

Since the determinant of the KdS metric (g) equals $-\rho^4 \sin^2\vartheta$, we obtain

$$\det(\overline{g}) = -\frac{c^2\rho^4}{E^2}(r_E\vartheta_K - r_K\vartheta_E)^2 \sin^2\vartheta. \tag{32}$$

From the concept of Gaussian coordinate systems, as developed, e.g., in Ref. [6], we know that a coordinate system based on geodesics that starts off with zero velocity has a metric with $\overline{g}_{0i} = \overline{g}_{i0} = 0, i = 1,2,3$ for as long as the metric is valid. Hence, our consideration of orbits with $L_z = 0$ only. That this is true in our case can be verified with (30) and $\mathcal{T} = C$: inserting the general equations of motion (with $L_z \neq 0$) into the expression for $2\overline{g}_{03}$ (factor of the term in $d\tau d\psi$), we find indeed that $\overline{g}_{03} = 0$ globally only if $L_z = 0$. For the coefficients $2\overline{g}_{01}$ (factor of the term in $d\tau dE$) and $2\overline{g}_{02}$ (factor of the term in $d\tau dK$) we need to make use of (22).

A zero 'start' velocity also means a zero velocity in the meridional plane. The radius at which the radial velocity is zero is an apocenter radius and we denote it with a subscript a, hence r_a. The polar angle ϑ at which the tangential velocity in the meridional plane is zero we similarly denote by ϑ_a. When the expressions for $R(r)$ and $\Theta(\vartheta)$, as given by (12) and (13), are equated to zero (as is the case for $r = r_a$ and $\vartheta = \vartheta_a$), they can easily be solved for E^2 and K:

$$\begin{aligned} -g_{33}(r_a, \vartheta_a)\rho^2(r_a, \vartheta_a)E^2 &= \sin^2\vartheta_a\rho^2(r_a, \vartheta_a)\Delta_r(r_a)\Delta_\vartheta(\vartheta_a) \\ -g_{33}(r_a, \vartheta_a)\rho^2(r_a, \vartheta_a)K &= \sin^2\vartheta_a\left[\left(r_a^2 + a^2\right)^2\cos^2\vartheta_a\Delta_\vartheta(\vartheta_a) + r_a^2\sin^2\vartheta_a\Delta_r(r_a)\right]. \end{aligned} \tag{33}$$

Clearly E^2 and K are positive for all initial conditions r_a and ϑ_a outside the black hole. Better still, r_a and ϑ_a are valid substitutes for E and K, with the advantage that they are intuitive (we assume $r_a < +\infty$).

The inequality (12) can also be written as $E^2 \geq V_{\text{eff}}(r)$ (with)

$$V_{\text{eff}}(r) = \frac{(r^2 + a^2K)\Delta_r(r)}{(r^2 + a^2)^2}. \tag{34}$$

A qualitative inspection of this curve shows that the orbits we consider here can also attain a radius where the radial velocity is again zero (hence a turning point) which we call the pericenter radius; we denote that radius by $\overline{r}_p$. In addition, there are orbits that can attain $r = 0$ with nonzero velocity. They are called polar orbits.

As for the 'orbital' part of the metric $\overline{g}_{ij}(\tau, E, K), i = 1,2,3$, we thus assume that at $t = \tau = 0$ all orbits start off with zero velocity. In that case, all the partial derivatives that appear in the transformation matrix (29) are regular in the vicinity of (r_a, ϑ_a), and hence the metric $\overline{g}_{ij}(\tau, E, K)$, as calculated in (30) with $\mathcal{T} = C$, is regular there. We also note that we deal here with very special orbits for which (r_a, ϑ_a) is a point of the orbit. In general r_a and ϑ_a will be reached at different times, since solving (33) does not imply that $\vartheta_a(\tau) = r_a(\tau)$ for some τ.

Now that we have established regularity of the metric $\overline{g}_{\alpha,\beta}$ for τ small, it still remains to be seen what happens at later times, since there is no theorem that guarantees regularity, and thus validity. Therefore, we have to go through the chore of actually calculating the $\overline{g}_{ij}(\tau, E, K), i = 1,2,3$. Since we are now in the process of getting rid of the singular coordinates t and φ, we can expect to also get rid of the singularities in the original KdS metric. The coefficients in the third row and/or column are

$$\overline{g}_{13} = \overline{g}_{31} = a_1 r_E + b_1 \vartheta_E + g_{33} w_E \tag{35}$$

$$\overline{g}_{23} = \overline{g}_{32} = a_1 r_K + b_1 \vartheta_K + g_{33} w_K \tag{36}$$

$$\overline{g}_{33} = g_{33} \tag{37}$$

with

$$\begin{aligned} a_1 &= \frac{1}{E\sqrt{R}}\left[(\overline{r}^2 + a^2)aE^2 \sin^2\overline{\vartheta} + (\overline{r}^2 + a^2K)g_{03}\right] \\ b_1 &= \frac{1}{E\sqrt{\Theta}}\left[(\overline{r}^2 + a^2)aE^2 \sin^2\overline{\vartheta} + a^2(K - \cos^2\vartheta)g_{03}\right]. \end{aligned} \tag{38}$$

The remaining coefficients are

$$\begin{aligned} \overline{g}_{21} = \overline{g}_{12} = {} & a_2 r_E r_K + b_2 \vartheta_E \vartheta_K + c_2(r_E \vartheta_K + r_K \vartheta_E) + \\ & + a_1(r_E w_K + r_K w_E) + b_1(\vartheta_E w_K + \vartheta_K w_E) + g_{33} w_E w_K \end{aligned} \tag{39}$$

$$\overline{g}_{11} = a_2 r_E^2 + b_2 \vartheta_E^2 + 2c_2 r_E \vartheta_E + 2a_1 r_E w_E + 2b_1 \vartheta_E w_E + g_{33} w_E^2 \tag{40}$$

$$\overline{g}_{22} = a_2 r_K^2 + b_2 \vartheta_K^2 + 2c_2 r_K \vartheta_K + 2a_1 r_K w_K + 2b_1 \vartheta_K w_K + g_{33} w_K^2 \tag{41}$$

with

$$\begin{aligned} a_2 &= \frac{\overline{r}^2 + a^2K}{E^2\overline{R}}\left[(\overline{r}^2 + a^2K)g_{00} - E^2\overline{\rho}'^2\right] \\ b_2 &= \frac{a^2(K - \cos^2\overline{\vartheta})}{E^2\overline{\Theta}}\left[a^2(K - \cos^2\overline{\vartheta})g_{00} - E^2\overline{\rho}'^2\right] \\ c_2 &= \frac{a^2}{E^2\sqrt{\overline{R}}\sqrt{\overline{\Theta}}}\left[(\overline{r}^2 + a^2K)(K - \cos^2\overline{\vartheta})g_{00} - E^2\left(2\overline{r}^2 \sin^2\overline{\vartheta} - \overline{\rho}^2 + K\overline{\rho}'^2\right)\right]. \end{aligned} \tag{42}$$

In the above expressions (35)–(39),[3] it is, of course, understood that $g_{00}(\overline{r}, \overline{\vartheta})$, $g_{03}(\overline{r}, \overline{\vartheta})$, and $g_{33}(\overline{r}, \overline{\vartheta})$. We also note that the metric is independent of ε_r and ε_ϑ.

---

[3] These expressions may look ugly, but considering that they hide considerable algebraic manipulation and a kind of 'magic' associated with the disappearance of the singularities in Δ_r and Δ_ϑ, their structure is actually quite elegant. Apart from the metric coefficients (g_{00}, g_{03}, g_{33}) and the inevitable partial derivatives which are irreducible at this stage without solving the equations of motion, they contain only 5 additional functions $(a_1, a_2, b_1, b_2, c_2)$ that do not depend on these partial derivatives.

Clearly, all singularities associated with Δ_r and Δ_ϑ are gone, but an $\overline{R}$ or a $\overline{\Theta}$ in a denominator causes a singularity whenever r_a, r_p or ϑ_a is reached. Technically, this is a consequence of having to split up a regular limit into two singular terms in the expressions for φ_E and φ_K. Hence, we have again two complementary representations of $\overline{g}_{\alpha\beta}$: the one in (29) with $\mathcal{T} = C$ that is valid in the vicinity of $\tau = 0$ and the one that is valid for each orbit (E, K) up to the time τ_p. Since (1) the orbits are continuously replenished from far away and (2) the singularities associated with the $R(\overline{r}(0, E, K))$ and $\Theta(\overline{\vartheta}(0, E, K))$ are removable, the metric given by Eqs. (35)–(39) suffices as the only metric.

The KdS metric is a regular one outside the ergosphere $r = M + \sqrt{M^2 - a^2 \cos^2\vartheta}$. The synchronous metric can be seen there as a kind of travelator or conveyor belt that one can 'step on' (this requires a boost to match the velocity of the travelator ring that passes by) to enter this valid metric that carries one through the ergosphere and the region $\Delta_r(r) < 0$ between the event horizons $R_\pm = M \pm \sqrt{M^2 - a^2}$. The ergosphere has disappeared, illustrating the fact that the presence of an ergosphere is a property of the velocity of the observer rather than an intrinsic feature of the rotating black hole. The travelator consists of rings labeled by (E, K), and these are centered on the polar axis $\vartheta = 0$. The travelator metric can be freely traveled as long as one stays clear of the τ_p of the rings that one is standing on. Since all $\overline{r}_p$ are located in the inner KdS spacetime (defined here as the spacetime inside the inner event horizon R_-), it is always possible to 'step off' the travelator 'in time' after having passed R_-, again by applying a boost to match the velocity of objects that are at rest in another valid metric.

As explained above, we need not bother with the singularities associated with R and Θ, in the KdS case, that is. However, in the next section, we will consider nonempty rotating black holes, and we will have to deal with these singularities. Inspection of the expressions (35)–(42) and those for w_E and w_K, in combination with the appropriate differentials dE and dK of the travelator metric, reveals that all singularities appear as

$$\frac{r_E}{\sqrt{R}} dE = 2\frac{d\sqrt{R}}{\partial_r R}, \qquad \frac{\vartheta_E}{\sqrt{\Theta}} dE = 2\frac{d\sqrt{\Theta}}{\partial_\vartheta \Theta} \tag{43}$$

and similarly for the differentials in K. Since the zeroes of R and Θ are simple zeroes, $\partial_r R$ and $\partial_\vartheta \Theta$ are nonzero there. Hence, if we pass, locally, from (E, K) to $(\sqrt{R}, \sqrt{\Theta})$, the metric in these coordinates is regular and Lorentzian.

It may seem remarkable that the travelator metric (35)–(42) has such a simple determinant (32). However, this can be verified using the common techniques of determinant evaluation, not without effort.

4.2 The case $a = 0$

The transformation matrix from the coordinates $(x^\alpha) = (t, r, \vartheta, \varphi)$ to the new coordinates $(\overline{x}^\alpha) = (\tau, E, \vartheta, \varphi)$ is now simply

$$(C_{\alpha\beta}) = \begin{pmatrix} t_\tau & r_\tau & 0 & 0 \\ t_E & r_E & 0 & 0 \\ 0 & 0 & 1 & 0 \\ 0 & 0 & 0 & 1 \end{pmatrix} = \left(\frac{\partial x^\beta}{\partial \overline{x}^\alpha}\right), \qquad \det(C) = \frac{c r_E}{E}. \tag{44}$$

Since ϑ and φ need no transformation, we obtain immediately from the KdS metric and the expression for a_2

$$ds^2 = c^2 d\tau^2 - \frac{r_E^2}{E^2} dE^2 - \bar{r}^2 d\vartheta^2 - \bar{r}^2 \sin^2\vartheta d\psi^2. \tag{45}$$

Note that in this metric $\bar{r}(E, \tau)$ as well as $r_E(E, \tau)$. This is the travelator version of the Schwarzschild-Λ metric (also called the Schwarzschild-de Sitter metric). In Ref. [1], this metric is called the Novikov metric.

4.3 The case $\vartheta = 0$

Here we are dealing with a subset of the orbits, those that are confined to the polar axis and that are therefore dynamically decoupled from the rest. The transformation matrix from the coordinates $(x^\alpha) = (t, r)$ to the new coordinates $(\bar{x}^\alpha) = (\tau, E)$ is the lower dimensional version of (44):

$$(C_{\alpha\beta}) = \begin{pmatrix} t_\tau & r_\tau \\ t_E & r_E \end{pmatrix} = \left(\frac{\partial x^\beta}{\partial \bar{x}^\alpha}\right), \qquad \det(C) = \frac{c r_E}{E}. \tag{46}$$

The coordinate φ is not present, and $K = 1$. Hence we obtain

$$ds^2 = c^2 d\tau^2 - \frac{r_E^2}{E^2} dE^2. \tag{47}$$

Again $r_E(E, \tau)$. The similarity with the previous case is only formal and qualitative, since here a is not necessarily zero and, therefore, the orbits are different. The result (47) can, of course, also be obtained from (35)–(43) with $d\vartheta = d\psi = 0$.

5. The Einstein equations

The KdS metric satisfies the Einstein equations

$$R^{\alpha\beta} - \frac{1}{2} R g^{\alpha\beta} - \Lambda g^{\alpha\beta} = 0 \tag{48}$$

with $R^{\alpha\beta}$ the Ricci tensor and R the Riemann scalar,[4] and that is, of course, no different for the travelator metric. From now on, we will lift the restriction that all orbits on which the travelator metric is built start off with zero velocity at a given time, say $\tau = 0$. The only consequence is that the travelator metric will also be singular at $\tau = 0$, but, as already indicated, these singularities are removable. Hence, any family of orbits that satisfies (15) and (16) gives rise to a travelator metric.

What we will need for the Einstein equations in the KdS metric can be found in Appendix A. The contravariant Ricci tensor $R^{\alpha\beta}$ in the KdS metric transforms into the travelator metric according to $\overline{R}^{\alpha\beta} = \frac{\partial \bar{x}^\alpha}{\partial x^\rho} \frac{\partial \bar{x}^\beta}{\partial x^\sigma} R^{\rho\sigma}$, or in matrix form

[4] Not to be confused with the R in 12, of course.

$$(\overline{R}) = \mathcal{T}^T(R)\mathcal{T} \tag{49}$$

with $\mathcal{T} = D = C^{-1}$. The matrix D is given in Appendix B. For as far as the contravariant metric tensor in the travelator metric is concerned, we will here only need the rather trivial $\overline{g}^{00} = 1/c^2$ and the $\overline{R}^{00}$ component of the contravariant Ricci tensor in Appendix C.

One could set out to verify (48) analytically for both the KdS and the travelator metric, but that is, again, a chore because all the derivatives of g_{00}, in addition to Δ_r and Δ_ϑ, must be exposed explicitly in terms of M, a and Λ.[5] For this paper, a numerical check has been performed, which is comparatively easy, and zeroes have been obtained up to machine accuracy. This latter fact shows, by the way, that the zeroes are not trivial but a result of all kinds of cancelations. Moreover, the identity (48) is satisfied irrespective of whatever values for r_E, ϑ_E, r_K, ϑ_K, w_E, and are assumed.

Next, we note that in the KdS metric the constants M and a are independent. Therefore, we can choose two new parameters, say λ and μ, unrelated to the KdS coordinates, and consider.[6]

$$M = m(\lambda, \mu) \quad \text{and} \quad a = A(\lambda, \mu). \tag{50}$$

We write the KdS metric with these functions instead of M and a. Looking now at the equations for the orbits (15)–(18), we see that all four constants M, a, E, and K can be freely chosen, which is a direct consequence of the separability of the equations of the motion. Hence we can also consider $E(\lambda, \mu)$ and $K(\lambda, \mu)$. We now transform the KdS metric, this time in its (λ, μ) livery, to the travelator metric (35)–(42), thus in terms of $m(\lambda, \mu)$, $A(\lambda, \mu)$, τ, $E(\lambda, \mu)$, $K(\lambda, \mu)$, ψ and the differentials $d\tau$, dE, dK, and $d\psi$. Next follows a transformation from (E, K) to the new variables (λ, μ), using the transformation (30) with

$$\mathcal{T} = U = \begin{pmatrix} 1 & 0 & 0 & 0 \\ 0 & E_\lambda & K_\lambda & 0 \\ 0 & E_\mu & K_\mu & 0 \\ 0 & 0 & 0 & 1 \end{pmatrix} \tag{51}$$

with obvious notations. That metric $(\overline{\overline{g}})$ (1) still satisfies the Einstein Eqs. (48) because of covariance, (2) features the variables τ, λ, μ, ψ and the differentials $d\tau$, $d\lambda$ (replacing dE), $d\mu$ (replacing dK) and $d\psi$ and (3) is the same as (34)–(41) with the subscripts E and K replaced by λ and μ.

We note that, since E and K are attributes of a comoving ring, λ and μ replace them as attributes of a comoving ring. We thus obtain a two-dimensional manifold of comoving rings, the motion of which is determined by (15) and (16), equations that feature $m(\lambda, \mu)$, $A(\lambda, \mu)$, $E(\lambda, \mu)$, and $K(\lambda, \mu)$.

If a and M are not functions of λ and μ, but just constants, then we have produced a travelator metric that is free of the KdS singularities and satisfies the Einstein

[5] In the course of the rather tedious calculations, we need all components of all the tensors involved. They are not listed here because of space limitations.

[6] The notation here is illogical for as far as the use of upper case and lower case is concerned, but is adopted in order to retain the rather established notations M and a.

equations for empty space (48). In that case, we have merely replaced the constants of the motion E and K by other constants of the motion λ and μ. But $A(\lambda,\mu)$ and $m(\lambda,\mu)$ need not be constants, and we now investigate what this means for the Einstein equations for a nonempty rotating black hole:

$$\overline{\overline{R}}^{\alpha\beta} - \frac{1}{2}R\overline{\overline{g}}^{\alpha\beta} - \Lambda\overline{\overline{g}}^{\alpha\beta} = \frac{8\pi G}{c^4}\overline{\overline{T}}^{\alpha\beta}, \tag{52}$$

with $\overline{\overline{T}}^{\alpha\beta}$ the stress-energy-momentum tensor.

We consider comoving matter:

$$\overline{\overline{u}}^{\alpha} = \left(\frac{d\tau}{d\tau}, \frac{d\lambda}{d\tau}, \frac{d\mu}{d\tau}, \frac{d\psi}{d\tau}\right) = (1,0,0,0). \tag{53}$$

The Ricci tensor $\overline{\overline{R}}^{\alpha\beta}$ contains the derivatives $\partial_r^n\Delta_r(r,M,a^2)$, $\partial_\vartheta^n\Delta_\vartheta(\vartheta,a^2)$ with $n=1,2$. When taken as such, i.e., with the partial derivative only pertaining to the first argument, all $\overline{\overline{T}}^{\alpha\beta}$ are zero. However, since λ and μ are now coordinates, the inverse functions $\lambda(\tau,r,\vartheta)$ and $\mu(\tau,r,\vartheta)$ also make sense. Indeed, the functions $\overline{r}(\tau,\lambda,\mu)$ and $\overline{\vartheta}(\tau,\lambda,\mu)$ from Section 3 tell us at which (r,ϑ) the ring with label (λ,μ) intersects the meridian plane at time τ. Hence, the inverse functions $\lambda(\tau,r,\vartheta)$ and $\mu(\tau,r,\vartheta)$ tell us which ring with label (λ,μ) happens to be at (r,ϑ) at time τ. We can, therefore, consider as alternative for $\partial_r\Delta_r$ the expression

$$\begin{aligned}\partial_r\Delta_r\left(r,m(\lambda,\mu),A^2(\lambda,\mu)\right) &= \partial_r\Delta_r\left(r,M,a^2\right) + m_{\overline{r}}\partial_m\Delta_r\left(r,m,A^2\right) + A^2_{\overline{r}}\partial_{A^2}\Delta_r\left(r,m,A^2\right)\\ &= \partial_r\Delta_r\left(r,M,a^2\right) - 2rm_{\overline{r}} + \left(1-\frac{\Lambda}{3}r^2\right)A^2_{\overline{r}}\\ &\equiv \partial_r\Delta_r\left(r,M,a^2\right) + \overline{\partial_r\Delta_r}\end{aligned} \tag{54}$$

and similarly for $\partial_\vartheta\Delta_\vartheta$

$$\begin{aligned}\partial_\vartheta\Delta_\vartheta\left(\vartheta,A^2(\lambda,\mu)\right) &= \partial_\vartheta\Delta_\vartheta\left(\vartheta,a^2\right) + A^2_{\overline{\vartheta}}\partial_{A^2}\Delta_\vartheta\left(\vartheta,A^2\right)\\ &= \partial_\vartheta\Delta_\vartheta\left(\vartheta,a^2\right) + \frac{1}{3}\Lambda\cos^2\vartheta A^2_{\overline{\vartheta}}\\ &\equiv \partial_\vartheta\Delta_\vartheta\left(\vartheta,a^2\right) + \overline{\partial_\vartheta\Delta_\vartheta}.\end{aligned} \tag{55}$$

The symbols with subscripts $\overline{r}$ and $\overline{\vartheta}$ stand for the expressions

$$\begin{aligned}f_{\overline{r}}(\tau,r,\vartheta) &= \lambda_r\partial_\lambda f + \mu_r\partial_\mu f, \qquad f = m, A^2\\ A^2_{\overline{\vartheta}}(\tau,r,\vartheta) &= \lambda_\vartheta\partial_\lambda A^2 + \mu_\vartheta\partial_\mu A^2\end{aligned} \tag{56}$$

and λ_r, λ_ϑ, μ_r, and μ_ϑ stand for partial derivatives of the functions $\lambda(\tau,r,\vartheta)$ and $\mu(\tau,r,\vartheta)$ in our usual notations. Since we rather have $\overline{r}(\tau,\lambda,\mu)$ and $\overline{\vartheta}(\tau,\lambda,\mu)$ at our disposition, these partial derivatives are calculated using

$$\begin{pmatrix}\lambda_r & \mu_r\\ \lambda_\vartheta & \mu_\vartheta\end{pmatrix}(\tau,\lambda,\mu) = \frac{1}{\overline{r}_\lambda\overline{\vartheta}_\mu - \overline{\vartheta}_\lambda\overline{r}_\mu}\begin{pmatrix}\overline{\vartheta}_\mu & -\overline{\vartheta}_\lambda\\ -\overline{r}_\mu & \overline{r}_\lambda\end{pmatrix}(\tau,\lambda,\mu). \tag{57}$$

Expressions for the second derivatives are obtained analogously to (54) and (55). We write

$$\begin{aligned}\partial_r^2\Delta_r\left(r,m(\lambda,\mu),A^2(\lambda,\mu)\right) &= \partial_r^2\Delta_r\left(r,M,a^2\right)+\overline{\partial_r^2\Delta_r}\\ \partial_\vartheta^2\Delta_\vartheta\left(\vartheta,A^2(\lambda,\mu)\right) &= \partial_\vartheta^2\Delta_\vartheta\left(\vartheta,a^2\right)+\overline{\partial_\vartheta^2\Delta_\vartheta}\end{aligned} \tag{58}$$

with again

$$\begin{aligned}\overline{\partial_r^2\Delta_r} &= \left(\lambda_r\partial_\lambda+\mu_r\partial_\mu\right)\left[\partial_r\Delta_r|_{M=m,a^2=A^2}+\overline{\partial_r\Delta_r}\right]\\ \overline{\partial_\vartheta^2\Delta_\vartheta} &= \left(\lambda_\vartheta\partial_\lambda+\mu_\vartheta\partial_\mu\right)\left[\partial_\vartheta\Delta_\vartheta|_{a^2=A^2}+\overline{\partial_\vartheta\Delta_\vartheta}\right].\end{aligned} \tag{59}$$

Whether one holds on to 'the original' $\partial_r^n\Delta_r$ and $\partial_\vartheta^n\Delta_\vartheta$ (in which M and a^2 are constants) or adopts the alternatives $\overline{\partial_r^n\Delta_r}$ and $\overline{\partial_\vartheta^n\Delta_\vartheta}$ (in which m and A^2 are functions of λ and μ), is a matter of choice for every equation (α,β) in (52) separately.

6. The Lemaitre–Tolman metric (the case $a = 0$)

We now illustrate the procedure that is outlined in the previous section for the simplest possible case $a = 0$. We need only the coordinate λ.

We first revisit the metric (45) in $(\tau,\lambda,\vartheta,\psi)$ (coordinates)

$$ds^2 = c^2d\tau^2 - \frac{\overline{r}_\lambda^2}{E^2(\lambda)}d\lambda^2 - \overline{r}^2d\vartheta^2 - \overline{r}^2\sin^2\vartheta d\psi^2, \tag{60}$$

with, again $\overline{r}_\lambda = \partial_\lambda\overline{r}$.[7] The metric (60) is exactly what has been called in Ref. [1] the Lemaitre-Tolman metric (hereafter L-T metric) if we set

$$E^2(\lambda) = 1 - 2e(\lambda) \tag{61}$$

in the notations of Ref. [1]. What we call here λ is mostly known as r in the L-T models, which was called the shell label in Ref. [1]. The only parameter M, which does not appear in the metric but solely in the equation of the motion (26), is now a function of λ. As before, we write $m(\lambda)$, and in this simple case we know that $m(\lambda)$ is the mass inside the shell with label λ. We assume $0 \le \lambda \le 1$ with the $\lambda = 1$ shell the shell beyond which there is no mass anymore.

As for the Einstein equations, we must first transform the contravariant tensors in the appendices which are in the (τ,E,ϑ,ψ) coordinates to the $(\tau,\lambda,\vartheta,\psi)$ coordinates with the transformation (49) using the transformation matrix $V = U^{-1}$. This

[7] This result is immediate from elementary calculus, but can, of course, also be obtained by considering the transformation 30 with

$$\mathcal{T} = U = \begin{pmatrix}1 & 0 & 0 & 0\\ 0 & E_\lambda & 0 & 0\\ 0 & 0 & 1 & 0\\ 0 & 0 & 0 & 1\end{pmatrix}. \tag{61}$$

This matrix transforms a covariant tensor of rang 2 into a tensor with the same coefficients, except for these that carry just one index 2, which are multiplied by E_λ, and the 22-index, which is multiplied by E_λ^2.

transformation produces a contravariant tensor with the same coefficients, except for these that carry just one index 2 which are divided by E_λ, and the 22-index, which is divided by E_λ^2. Since the Einstein equations feature only diagonal coefficients, this transformation has no effect on the final equations.

The stress-energy-momentum tensor reads, in spherical geometry with anisotropic pressures (see, e.g., Doneva & Yazadjiev [7]):

$$\overline{\overline{T}}^{\alpha\beta} = \rho_0 \overline{\overline{u}}^{\alpha} \overline{\overline{u}}^{\beta} + p_r e_r^\alpha e_r^\beta + p_t \left(\overline{\overline{u}}^{\alpha} \overline{\overline{u}}^{\beta} - e_r^\alpha e_r^\beta - g^{\alpha\beta} \right), \tag{62}$$

since the λ-direction is the radial direction. The tensor that an arbitrary tangential pressure p_t (i.e., a pressure orthogonal to the radial direction) is multiplied by in (63), is orthogonal to both $\overline{\overline{u}}^{\alpha}$ and e_r^α.

We find for the 00-equation, with the expressions in Appendices A and C.2.

$$-\frac{1}{c^2}\left(\frac{\Delta - 1}{r^2} + \frac{\partial_r \Delta}{r} + \Lambda \right) = \frac{8\pi G}{c^4} \overline{\overline{T}}^{00}. \tag{63}$$

When M is a constant, this expression reduces to $0 = 0$. When $m(\lambda)$, we obtain

$$-\frac{\overline{\partial_r \Delta_r}}{\overline{r}^3} = \frac{8\pi G}{c^2} \overline{\overline{T}}^{00}, \tag{64}$$

or, with (54)

$$2\frac{\lambda_r \partial_\lambda m}{\overline{r}^2} = \frac{8\pi G}{c^2} \overline{\overline{T}}^{00}. \tag{65}$$

For a comoving medium we have $\overline{\overline{T}}^{00} = \overline{T}^{00} = \rho_0$ and in this simple case $\lambda_r^{-1} = \overline{r}_\lambda$, thus

$$\frac{\partial_\lambda m(\lambda)}{\overline{r}^2(\lambda, \tau) \overline{r}_\lambda(\lambda, \tau)} = \frac{4\pi G}{c^2} \rho_0. \tag{66}$$

This is exactly the same equation (mutatis mutandis) as the one valid for the L–T models, as, e.g., given in Eq. (6) of Ref. [1]. It is the general-relativistic form of Poisson's equation.

The 11-equation is almost identical to the 00-equation and yields again either $0 = 0$, (or)

$$\frac{\partial_\lambda m}{\overline{r}^2 \overline{r}_\lambda} = -\frac{\overline{r}_\lambda^2}{E^2} \frac{4\pi G}{c^4} \overline{\overline{T}}^{11}. \tag{67}$$

Since $\overline{\overline{T}}^{11} = -p_r \overline{\overline{g}}^{11}$ is a positive quantity or zero, this equation is not compatible with (66), unless we choose for this equation $\partial_\lambda m = 0$. This shows that we are free to make this choice for each equation, since we obtained already the L-T solution, and this choice confirms this freedom. We conclude that the flux of the radial flow through any sphere is zero, confirming that spheres are comoving surfaces.

The 22-equation equals

$$\frac{1}{2r^2}\partial_r^2\Delta + \frac{1}{r^3}\partial_r\Delta + \frac{1}{r^2}\Lambda = \frac{8\pi G}{c^4}\overline{\overline{T}}^{22} \tag{68}$$

Here $\overline{\overline{T}}^{22} = T^{22} = -p_t\overline{\overline{g}}^{22}$. Again, $\overline{\overline{T}}^{22} = 0$ is the solution if $\partial_\lambda m = 0$. If $\partial_\lambda m \neq 0$ we obtain

$$-\frac{1}{\overline{r}\overline{r}_\lambda^2}\partial_\lambda^2 m = \frac{8\pi G}{c^4}\overline{\overline{T}}^{22}. \tag{69}$$

We know from Ref. [1] that regular solutions in the center behave like $m(\lambda) \sim \lambda^3$ for $\lambda \to 0$. Hence we conclude that a $\overline{\overline{T}}^{22} \geq 0$ is impossible, unless we allow for a singularity in the center.

The 33-equation adds no further information, since $\sin^2\vartheta\overline{\overline{R}}^{33} = \overline{\overline{R}}^{22}$, $\sin^2\vartheta\overline{\overline{g}}^{33} = \overline{\overline{g}}^{22}$ and $\sin^2\vartheta\overline{\overline{T}}^{33} = \overline{\overline{T}}^{22}$.

In this simple case, we have three independent equations, yielding two structurally different solutions: (1) empty space with central singularity, i.e., the Schwarzschild-Λ solution and (2) the L-T solution.

In order to make the L-T solution explicit, we still need to prescribe $0 \leq e(\lambda) \leq \frac{1}{2}$ and $0 \leq m(\lambda) \leq M = m(1)$. The latter inequality follows from the requirement that the outermost massive shell $\lambda = 1$ must match the Schwarzschild-Λ solution at maximum expansion. The functions $e(\lambda)$ and $m(\lambda)$ must be such that no shell $\lambda < 1$ has a maximum expansion that is larger than the outermost $\lambda = 1$ shell. We also need to assign, for a given cosmic time τ and for every shell λ, a function $r(\lambda)$, together with the specification whether the shell is expanding or contracting. In the parlance of Ref. [1] this means the specification of a phase function.

Out of the great variety of models that come with these choices, there is a special class of models for which all shells attain their maximum extent at the same cosmic time. These shells will have in general different radial periods which will cause eventually shell collisions. The radial period of shell λ is given by (24):

$$2T_r = 2\int_0^{\overline{r}_0} \frac{\sqrt{r}dr}{\sqrt{2m(\lambda) - 2e(\lambda)r + \Lambda r^3/3}}, \tag{70}$$

with $\overline{r}_0$ the smallest positive root of the radicand. Writing, as in Ref. [1],

$$\overline{r} = p(\lambda)\,\mathrm{cyc}(e, m, \Lambda, \tau) \tag{71}$$

we obtain

$$\frac{p^{3/2}}{\sqrt{2m}}\int_0^{w_0} \frac{\sqrt{\mathrm{cyc}}d\mathrm{cyc}}{\sqrt{1 - (ep/m)\mathrm{cyc} + (\Lambda/6)(p^3/m)\mathrm{cyc}^3/3}}, \tag{72}$$

with w_0 the smallest real positive root for cyc of the radicand. If that radial period is to be the same for all shells, which means that shells will not cross, the radicand should not depend on λ. Assuming $p(\lambda) = 2M\lambda$, we find that $e(\lambda) = \frac{1}{2}\lambda^2$ and $m(\lambda) = M\lambda^3$. This is the spherical version of the Robertson–Walker metric for a closed universe.

Finally, we remark that we did obtain the L-T solution starting from the Schwarzschild-Λ solution (since $a = 0$), 'simply' by choosing another coordinate system that is comoving. This shows that the L–T solution can be derived from the Schwarzschild-Λ solution by coordinate transformation, but that could only be accomplished because the equations of motion are separable.

7. Dusty models in the case $a \neq 0$

7.1 The general case

We now include a (pressureless) dusty medium with comoving proper-density ρ_0 in the travelator metric; the case $a = 0$ has learned us that pressures are unlikely to be realistic within this paradigm. The stress-energy-momentum tensor, therefore, has the simplest nonzero form possible: $\overline{\overline{T}}^{00} = \rho_0$ and all other components are zero. We note that, while in many cases, the Eq. (52) could also be taken in its covariant form, this is not the case; because the metric $\overline{\overline{g}}^{\alpha\beta}$ is not diagonal, it is imperative to use the contravariant form for the stress-energy-momentum tensor. All 10 equations of (52) remain obviously satisfied if $\rho_0 = 0$. We will now investigate which special form ρ_0 has to assume in order to satisfy

$$\overline{\overline{R}}^{00} - \frac{1}{2}\overline{\overline{R}}\,\overline{\overline{g}}^{00} - \Lambda\overline{\overline{g}}^{00} = \frac{8\pi G}{c^4}\rho_0 \tag{73}$$

if $\rho_0 \neq 0$.
With the aid of Appendix C we find

$$\begin{aligned}
&\frac{a^2}{2\overline{\rho}^4}\left(K - \cos^2\overline{\vartheta}\right)\partial^2_{\overline{r}}\Delta_r - \frac{\overline{r}}{\overline{\rho}^6}\left[\overline{\rho}^2 + 2a^2\left(K - \cos^2\overline{\vartheta}\right)\right]\partial_{\overline{r}}\Delta_r - 12\overline{\rho}^4\left(\overline{r}^2 + a^2K\right)\partial^2_{\overline{\vartheta}}\Delta_\vartheta - \\
&-\left\{\frac{a^2\left(K - \cos^2\overline{\vartheta}\right)}{\overline{\rho}^4}\left[3\left(\overline{r}^2 + a^2\right) + a^2\sin^2\overline{\vartheta}\right] + \frac{2a^2\sin^2\overline{\vartheta}}{\overline{\rho}^2} + 3\right\}\frac{\cos\overline{\vartheta}}{2\overline{\rho}^2}\frac{\partial_{\overline{\vartheta}}\Delta_\vartheta}{\sin\overline{\vartheta}} - \\
&-\frac{g_{00}}{\overline{\rho}^4}\left[\left(\overline{r}^2 + a^2K\right)\left(1 - \frac{4\overline{r}^2}{\overline{\rho}^2}\right) + 2\overline{r}^2\right] + \Delta_\vartheta\overline{\rho}^2 + \\
&+\frac{a^2\left(g_{00} - \Delta_\vartheta\right)}{\overline{\rho}^4}\left[\left(K - \cos^2\overline{\vartheta}\right)\left(\frac{4a^2\cos^2\overline{\vartheta}}{\overline{\rho}^2} - 1\right) + 2\cos^2\overline{\vartheta}\right] - \Lambda = \frac{8\pi G}{c^2}\rho_0.
\end{aligned}$$

The right-hand side of this expression evaluates to 0 if M and a^2 are constants. If, on the other hand, we consider the alternatives (54), (55), and (58), the additional terms that these dependencies entail are those featuring the derivatives of the Δ's:

$$\begin{aligned}
\frac{8\pi G}{c^2}\overline{\rho}_0 = &\frac{A^2}{2\rho^4}\left(K - \cos^2\vartheta\right)\overline{\partial^2_r\Delta_r} r\rho^6\left[\rho^2 + 2A^2\left(K - \cos^2\vartheta\right)\right]\overline{\partial_r\Delta_r}12\rho^4\left(r^2 + A^2K\right)\overline{\partial^2_\vartheta\Delta_\vartheta} + \\
&+\left\{\frac{A^2(1-K)}{\rho^4}\left[3\left(r^2 + A^2\right) + A^2\sin^2\vartheta\right] - A^2\sin^2\vartheta\rho^4\left[5r^2 + 4A^2 + A^2\cos^2\vartheta\right] - 3\right\}\frac{\cos\vartheta}{2\rho^2}\frac{\overline{\partial_\vartheta\Delta_\vartheta}}{\sin\vartheta},
\end{aligned}$$

resulting in an expression for the comoving density ρ_0. It is interesting to note that the 00 equation is the only Einstein equation that does not feature any of the r_E, r_K, ϑ_E, ϑ_K, w_E or w_K.

7.2 The case $\Lambda = 0$

We will now illustrate the analysis for the case $\Lambda = 0$. The expression for the comoving density simplifies greatly:

$$\frac{8\pi G}{c^2}\overline{\rho}_0 = \frac{A^2}{2\rho^4}\left(K - \cos^2\vartheta\right)\overline{\partial_r^2\Delta_r} - \frac{r}{\rho^6}\left[\rho^2 + 2A^2\left(K - \cos^2\vartheta\right)\right]\overline{\partial_r\Delta_r} \tag{74}$$

In order to make this solution more explicit, we need to prescribe $m(\lambda, \mu)$ and $A(\lambda, \mu)$ on the one hand, and $E(\lambda, \mu)$ and $K(\lambda, \mu)$ on the other hand. It is convenient to replace (E, K) by (r_a, ϑ_a), with now, of course, $r_a(\lambda, \mu)$ and $\vartheta_a(\lambda, \mu)$. The constants E and K are then given by (33). We will assume $0 \le \lambda \le 1$ and , with no constituent matter 'beyond' the $\lambda = 1$ rings. We recall that the comoving rings form a two-dimensional manifold, each ring being labeled by a pair (λ, μ), or a pair (E, K). From the case, $a = 0$ we also learned that each shell there moves as if it were in a Schwarzschild-Λ metric with mass $m(\lambda)$, independent of time. Likewise, in this case, each ring moves as if it were in a KdS metric with parameters $m(\lambda, \mu)$ and $A(\lambda, \mu)$, again, independent of time.

Since $\Theta(\vartheta) \ge 0$, every orbit is located in the region $\vartheta_a \le \vartheta \le \pi - \vartheta_a$, with the aperture ϑ_a given by $\Theta = -E^2 \sin^2\vartheta_a + K - \cos^2\vartheta_a = 0$. Since $R(r) \ge 0$, every orbit is located between r_a and $\overline{r}_p$, $\overline{r}_p \le r \le r_a$. There are no restrictions on $0 \le \vartheta_a \le \pi/2$.

As an example, we now assume $\sin^2\vartheta_a = \mu$. In order to avoid the KdS singularity at $r = 0$ and $\vartheta = \pi/2$, we assume $A(0, \mu) = 0$. The simplest choice seems to be

$$\begin{aligned} A^2(\lambda, \mu) &= \left(\lambda^2\mu\right)a^2 \quad \text{for} \quad \lambda \to 0 \\ &= a^2 - a_b(\mu)(1-\lambda)^\alpha \quad \text{for} \quad \lambda \to 1, \end{aligned} \tag{75}$$

with $a_b(\mu)$ a function and α a constant. The factor λ^2 ensures that $A^2 \to 0$ for $\lambda \to 0$ sufficiently smoothly.

As for $m(\lambda, \mu)$, we follow Ref. [1] and propose a monotonically increasing function such that

$$\begin{aligned} m(\lambda, \mu) &= m_c\lambda^3 \quad \text{for} \quad \lambda \to 0 \\ &= M - m_b(1-\lambda)^\beta \quad \text{for} \quad \lambda \to 1, \end{aligned} \tag{76}$$

with , m_b and β real constants. We know from the $a = 0$ case that the shell $E = 0$ must reach the Schwarzschild-Λ horizon M, and all other shells must satisfy $r_a \le M$. Likewise here, in order to realize the match between the inner and the outer KdS solution, at least one ring with $E = 0$ must have the apocenter radius $R_+ = M + \sqrt{M^2 - a^2}$. We will assume for simplicity that all rings $E = 0$ do.

We are still left with the specification of $r_a(\lambda, \mu)$. The apocenter radius r_a plays the same role (but is not the same) as the quantity $p(r)$ in Ref. [1]. We already indicated that $r_a(1, \mu) = R_+$. For simplicity, we assume

$$\begin{aligned} r_a(\lambda, \mu) &= r_{a,c}\lambda \quad \text{for} \quad \lambda \to 0 \\ &= R_+ - r_{a,b}(1-\lambda)^\gamma \quad \text{for} \quad \lambda \to 1, \end{aligned} \tag{77}$$

with $r_{a,c}$, $r_{a,b}$ and γ constants. The functions $\vartheta_a(\mu)$ and $r_a(\lambda)$ now being known, the energy follows from (33). We find

$$1-E^2 \equiv 2e \to \frac{2m_c}{r_{a,c}}\lambda^2 \quad \text{for} \quad \lambda \to 0$$
$$\to 1-\frac{m_b}{M}(1-\lambda)^\beta + \frac{r_{a,b}}{MR_+}\sqrt{M-a^2}(1-\lambda)^\gamma \qquad \text{for} \quad \lambda \to 1, \tag{78}$$

in agreement with Ref. [1] for $\lambda \to 0$.

In order to calculate $\rho(\tau, \lambda, \mu)$, we still have to choose initial conditions. This is no different from the case $a = 0$, where we have seen that the classical cosmological models (in their spherical form) also require very special initial conditions in order to avoid collisions of shells. Out of the orbits discussed in Appendix D we can choose those that all attain their apocenter radius (i.e., maximum expansion) at the same time τ. Numerical experiments show that it is indeed possible to choose initial conditions such that $\rho_0 > 0$ for subsequent times up to maximum expansion. This statement should be interpreted as no more than a proof of concept: i.e. a model for a rotating nonempty black hole or a rotating cosmological model with a travelator metric (35)–(42). A more detailed exploration of these models is beyond the scope of this paper.

8. Mass ejection or light emission

In Ref. [1] it has been shown that in general the g_{11} coefficient in the metric (69) will prevent not comoving matter and light to leave the black hole. More precisely, when an orbit (matter or light) attains the shell for which $E(\lambda) = 1 - 2e(\lambda) = 0$, the coefficient g_{11} will become in general singular, causing an outgoing orbit to turn back.[8] It is important that the condition $E(\lambda) = 0$ is independent of τ, hence it is the property of one particular shell to act as a barrier. That shell is the outermost shell that carries some constituent matter, and it is the only shell that can reach the Schwarzschild radius from within. Only when, at the very same time that the orbit reaches that shell $E = 0$, also (1) $\partial_\lambda \bar{r}(\lambda, \tau) = 0$ for that shell, which has been shown in Ref. [1] to be the condition that shells collide (in the continuum limit), and (2) that shell reaches the Schwarzschild radius, is it possible to create a 'neck', i.e., a nonzero 0/0 in g_{11}, through which matter or light can escape. It was noted in Ref. [1] that in spherical symmetry, these conditions are very unlikely to be globally satisfied, but that along the polar axis of a rotating black hole, there might be a possibility due to the similarity of the Kerr metric and the Schwarzschild metric along the polar axis. But that statement was a conjecture.

Now, we are in the position to be more precise with the help of the travelator metric (35)–(42). A (λ, μ) ring of constituent comoving matter orbits in a region bounded by 2 radii, $\bar{r}_p$ and r_a determined by the condition $R = 0$, and a cone $\vartheta_a \le \vartheta \le \pi - \vartheta_a$, with ϑ_a determined by the condition $\Theta = 0$. Every now and then, the ring will touch these boundaries, which act as impenetrable walls. At such an event, all coefficients of the orbital part of the travelator metric become singular, but, as already indicated, the singularities are removable locally. We note that at all other times, the comoving ring produces a regular metric.

Any particle, matter or light, that is not comoving will continuously visit different rings in (λ, μ) space. It might visit a comoving ring that just happens to touch its

[8] A particle or photon does not 'turn back' in the normal sense of the verb; it continues its course 'straight on'. It is the universe that 'flips over', as explained in Ref. [1].

boundary. Hence the particle will be forced to visit a ring for which $R > 0$ of $\Theta > 0$, that is, it remains in the interior of the black hole since all comoving rings remain in the interior of the black hole.

The metric along the polar axis (47), on the other hand, is fundamentally different, since no singularities associated with $R = 0$ or are present. That metric is formally identical to the L-T metric, and the mechanism as indicated above can take place. Hence, if the conditions are right, matter or light can leave the rotating black hole through the poles.

9. Concluding remark

The paradigm presented here can also be used to construct models for rotating dust balls, or rotating universes that are not the interior of a black hole. It suffices that all rings have $0 < E(\lambda, \mu) < 1$ and judicious choices for $A^2(\lambda, \mu)$ and $m(\lambda, \mu)$ are made. Since the mechanism outlined above for outflow along the polar axis is independent of the presence of a horizon, it opens up the possibility of bipolar outflows in fast-rotating dust balls, such as the vicinity of very young stars.

A. The Ricci tensor and Rieman scalar in the KdS spacetime

We adopt here the definition $R_{\alpha\beta} = \Gamma^{\rho}_{\alpha\beta,\rho} - \Gamma^{\rho}_{\alpha\rho,\beta} + \Gamma^{\rho}_{\sigma\rho}\Gamma^{\sigma}_{\alpha\beta} - \Gamma^{\rho}_{\sigma\beta}\Gamma^{\sigma}_{\alpha\rho}$. The tables that follow are to be read as follows: the quantity in the first column is the sum of terms consisting of the product of the second column with the third.

R	$\partial_r^2 \Delta_r$	$\frac{1}{\rho^2}$
	$\partial_\vartheta^2 \Delta_\vartheta$	$\frac{1}{\rho^2}$
	$\frac{\partial_\vartheta \Delta_\vartheta}{\sin\vartheta}$	$\frac{3\cos\vartheta}{\rho^2}$
	Δ_ϑ	$-\frac{2}{\rho^2}$

R^{00}	$\partial_r^2 g_{00} \frac{1}{2\rho^2\Delta_r}$	$(r^2 + a^2)^2$
	$\partial_\vartheta^2 g_{00} \frac{1}{2\rho^2\Delta_\vartheta}$	1
	$\partial_r g_{00}$	$-\frac{r}{\rho^2 \sin^2\vartheta} \frac{1}{\Delta_r\Delta_\vartheta} g_{33}$
	$\partial_\vartheta g_{00} \frac{\cot\vartheta}{2\rho^4}$	$\frac{2(r^2+a^2)^2}{\Delta_r} - \frac{\rho'^2}{\Delta_\vartheta}$

B. The transformation matrix D

All r and ϑ in this section should be read as $\bar{r}$ and $\bar{\vartheta}$.

B.1 The case $a \neq 0$

$$
\begin{aligned}
D_{00} &= \frac{E}{c} \\
D_{01} &= \frac{E}{\rho^2(r_E\vartheta_K - \vartheta_E r_K)}\left[\sqrt{R}\,\vartheta_K + \sqrt{\Theta}\,r_K\right] \\
D_{02} &= -\frac{E}{\rho^2(r_E\vartheta_K - \vartheta_E r_K)}\left[\sqrt{R}\,\vartheta_E + \sqrt{\Theta}\,r_E\right] \\
D_{03} &= \frac{E}{\rho^2(r_E\vartheta_K - \vartheta_E r_K)}\left[\sqrt{R}(\vartheta_E w_K - \vartheta_K w_E) + \sqrt{\Theta}(r_E w_K - r_K w_E)\right]
\end{aligned} \tag{79}
$$

$$
\begin{aligned}
D_{10} &= \frac{1}{c}\sqrt{R}\Delta_r \\
D_{11} &= \frac{1}{\rho^2(r_E\vartheta_K - \vartheta_E r_K)}\left[\frac{\sqrt{R}}{\Delta_r}\left(\sqrt{R}\,\vartheta_K + \sqrt{\Theta}\,r_K\right) + \rho^2\vartheta_K\right] \\
D_{12} &= -\frac{1}{\rho^2(r_E\vartheta_K - \vartheta_E r_K)}\left[\frac{\sqrt{R}}{\Delta_r}\left(\sqrt{R}\,\vartheta_E + \sqrt{\Theta}\,r_E\right) + \rho^2\vartheta_E\right] \\
D_{13} &= \frac{aE(r^2 + a^2)}{\sqrt{R}\Delta_r} + \left[\frac{R}{\Delta_r} + \rho^2\right]\frac{\vartheta_E w_K - \vartheta_K w_E}{\rho^2(r_E\vartheta_K - \vartheta_E r_K)} + \frac{\sqrt{R\Theta}}{\Delta_r}\frac{r_E w_K - r_K w_E}{\rho^2(r_E\vartheta_K - \vartheta_E r_K)}
\end{aligned} \tag{80}
$$

$$
\begin{aligned}
D_{20} &= -\frac{1}{c}\sqrt{\Theta}\Delta_\vartheta \\
D_{21} &= -\frac{1}{\rho^2(r_E\vartheta_K - \vartheta_E r_K)}\left[\frac{\sqrt{\Theta}}{\Delta_\vartheta}\left(\sqrt{R}\,\vartheta_K + \sqrt{\Theta}\,r_K\right) + \rho^2 r_K\right] \\
D_{22} &= \frac{1}{\rho^2(r_E\vartheta_K - \vartheta_E r_K)}\left[\frac{\sqrt{\Theta}}{\Delta_\vartheta}\left(\sqrt{R}\,\vartheta_E + \sqrt{\Theta}\,r_E\right) + \rho^2 r_E\right] \\
D_{23} &= \frac{aE}{\sqrt{\Theta}\Delta_\vartheta} - \left[\frac{\Theta}{\Delta_\vartheta} + \rho^2\right]\frac{r_E w_K - r_K w_E}{\rho^2(r_E\vartheta_K - \vartheta_E r_K)} - \frac{\sqrt{R\Theta}}{\Delta_\vartheta}\frac{\vartheta_E w_K - \vartheta_K w_E}{\rho^2(r_E\vartheta_K - \vartheta_E r_K)}
\end{aligned} \tag{81}
$$

$$
D_{30} = D_{31} = D_{32} = 0, \qquad D_{33} = 1. \tag{82}
$$

C. The contravariant Ricci tensor in the travelator spacetime

C.1 The case $a \neq 0$

All r and ϑ in this section should be read as $\bar{r}$ and $\bar{\vartheta}$.

$\overline{R}^{00}$	$\frac{1}{c^2}\frac{a^2}{2\rho^4}$	$\partial_\vartheta^2\Delta_\vartheta$	$-(K - \cos^2\vartheta)$
		$\frac{\partial_\vartheta\Delta_\vartheta}{\sin\vartheta}\frac{\cos\vartheta}{\rho^2}$	$-(K - \cos^2\vartheta)\left[3(r^2 + a^2) + a^2\sin^2\vartheta\right] - 2\rho^2\sin^2\vartheta$
		$2(g_{00} - \Delta_\vartheta)$	$(K - \cos^2\vartheta)\left(\frac{4a^2}{\rho^2}\cos^2\vartheta - 1\right) + 2\cos^2\vartheta$
	$\frac{1}{c^2}\frac{1}{\rho^4}$	$\frac{1}{2}\partial_r^2\Delta_r$	$r^2 + a^2K$
		$\partial_r\Delta_r\frac{r}{\rho^2}$	$-r^2 + a^2\cos^2\vartheta - 2a^2K$
		g_{00}	$(r^2 + a^2K)\left(\frac{4r^2}{\rho^2} - 1\right) - 2r^2$

C.2 The case $a = 0$

Taking account of

$$\begin{aligned}\Delta_\vartheta &= \vartheta_K = 1\\ R &= r^2\left(E^2r^2 - \Delta_r\right) = r^4\left(E^2r^2 - \Delta\right)\\ r_K &= \vartheta_E = w_E = w_K = \Theta = 0\\ g_{00} &= \Delta\end{aligned}$$

we find

$$\overline{R}^{00} = \frac{1}{c^2}\left(\frac{1}{2}\partial_r^2\Delta + \frac{1}{r}\partial_r\Delta\right) \tag{83}$$

D. The orbits in the case $\Lambda = 0$

D.1 The solution for $\overline{\vartheta}(r)$

In this appendix we first solve (15) explicitly. We denote

$$I_r(\overline{r}) = \int_{\overline{r}}^{\overline{r}_0} \frac{dr}{\sqrt{R(r)}} \tag{84}$$

and we need to solve

$$I_r(\overline{r}) = I_\vartheta\left(\overline{\vartheta}\right) = \int_{\overline{\vartheta}_0}^{\overline{\vartheta}} \frac{d\vartheta}{\sqrt{\Theta(\vartheta)}} = \frac{1}{a}\int_{\overline{\vartheta}_0}^{\overline{\vartheta}} \frac{d\vartheta}{\sqrt{K - E^2 - \left(1 - E^2\right)\cos^2\vartheta}}. \tag{85}$$

In this expression $\overline{\vartheta}_0$ is any attainable polar angle of the orbit. Be $0 \le \vartheta_a \le \pi/2$ the root of the radicand:

$$\cos\vartheta_a = k, \qquad k^2 = \frac{K - E^2}{1 - E^2} \le 1. \tag{86}$$

We now choose $\overline{\vartheta}_0 = \vartheta_a$, and thus $(\overline{r}_0, \vartheta_a)$ is a point along the orbit where the tangential velocity in the meridional plane is zero. We consider first the branch for increasing ϑ:

$$I_{\vartheta,1}\left(\overline{\vartheta}\right) = \int_{\vartheta_a}^{\overline{\vartheta}} \frac{d\vartheta}{\sqrt{\Theta(\vartheta)}} = \frac{1}{a}\int_{\vartheta_a}^{\overline{\vartheta}} \frac{d\vartheta}{\sqrt{K - E^2 - \left(1 - E^2\right)\cos^2\vartheta}}, \quad \vartheta_a \le \overline{\vartheta} \le \pi - \vartheta_a. \tag{87}$$

The change of variable

$$\nu = \frac{1}{k}\cos\vartheta, \qquad -1 \le \nu \le 1 \tag{88}$$

leads to the equation

$$a\sqrt{K-E^2}I_{\vartheta,1}(\overline{\nu}) = k\int_{\overline{\nu}}^{1}\frac{d\nu}{\sqrt{1-\nu^2}\sqrt{1-k^2\nu^2}}. \tag{89}$$

This can be rewritten as

$$\frac{\mathrm{sn}^{-1}}{}(\overline{\nu}) = \int_0^{\overline{\nu}} d\nu\sqrt{1-\nu^2}\sqrt{1-k^2\nu^2} = \boldsymbol{K}(k) - \frac{a}{k}\sqrt{K-E^2}I_{\vartheta,1}(\overline{\nu}), \tag{90}$$

with $\boldsymbol{K}(k)$ the complete elliptic integral of the first kind. Inversion produces

$$\cos\overline{\vartheta} = k\,\mathrm{sn}\left(\boldsymbol{K}(k) - \frac{a}{k}\sqrt{K-E^2}I_r(\overline{r})\right). \tag{91}$$

Clearly, the angular period $2\Delta I_\vartheta = 2\int_{\vartheta_a}^{\pi-\vartheta_a}$, defined as the increment of the argument of the elliptic sine after which sn repeats itself, is $4\boldsymbol{K}(k)$. When $I_r(\overline{r}) \to 0$ we obtain $\cos\overline{\vartheta} = k\,\mathrm{sn}(\boldsymbol{K}(k)) = k = \cos\vartheta_a$.

As for the second branch of the integral in ϑ, for decreasing ϑ, we write

$$I_{\vartheta,2}(\overline{\vartheta}) = \int_{\vartheta_a}^{\overline{\vartheta}}\frac{d\vartheta}{\sqrt{\Theta(\vartheta)}} = \int_{\vartheta_a}^{\pi-\vartheta_a}\frac{d\vartheta}{\sqrt{\Theta(\vartheta)}} + \int_{\pi-\vartheta_a}^{\overline{\vartheta}}\frac{d\vartheta}{\sqrt{\Theta(\vartheta)}}, \tag{92}$$

which can be written as

$$\frac{a}{k}\sqrt{K-E^2}I_{\vartheta,2}(\overline{\nu}) = 3\boldsymbol{K}(k) + \int_0^{\overline{\nu}}\frac{d\nu}{\sqrt{1-\nu^2}\sqrt{1-k^2\nu^2}}. \tag{93}$$

Inversion yields

$$\cos\overline{\vartheta} = k\,\mathrm{sn}\left(\frac{a}{k}\sqrt{K-E^2}I_r(\overline{r}) - 3\boldsymbol{K}(k)\right). \tag{94}$$

As for $I_r(\overline{r})$, we again need to consider two cases. When $\overline{r}$ decreases, we write $\int_{\overline{r}}^{\overline{r}_0} = \int_{\overline{r}}^{r_a} - \int_{\overline{r}_0}^{r_a}$. When $\overline{r}$ subsequently increases, we write $\int_{\overline{r}}^{\overline{r}_0} = \int_{\overline{r}_p}^{\overline{r}_0} + \int_{\overline{r}_p}^{\overline{r}} = \Delta I_r + \int_{\overline{r}_p}^{\overline{r}}$, and so on, such that $I_r(\overline{r})$ always increases. Hence, we are left with the integrals $\int_{\overline{r}}^{r_a}$ and $\int_{\overline{r}_p}^{\overline{r}}$.

The radicand R has four roots. One of them is $r_0 = r_a$. In case the other three are real we choose the largest of them as $r_1 = \overline{r}_p$, and the others as r_2 and r_3. In case two of them are complex conjugate, we choose the other real root as $r_1 = \overline{r}_p$, and r_2 and r_3 are the complex conjugate ones. The first integral we consider reads

$$I_{r,1}(\overline{r}) = \int_{\overline{r}}^{r_a}\frac{dr}{\sqrt{R(r)}} = \frac{1}{\sqrt{1-E^2}}\int_{\overline{r}}^{r_0}\frac{dr}{\sqrt{(r_0-r)(r-r_1)(r-r_2)(r-r_3)}}. \tag{95}$$

The transformation $t = (r-\overline{r})/(r_0-r)$ brings this integral in the form

$$I_{r,1}(\overline{r}) = \frac{2}{\sqrt{1-E^2}}\frac{\sqrt{r_0-\overline{r}}}{\sqrt{(r_0-r_1)(r_0-r_2)(r_0-r_3)}}R_F\left(\frac{\overline{r}-r_1}{r_0-r_1}, \frac{\overline{r}-r_2}{r_0-r_2}, \frac{\overline{r}-r_3}{r_0-r_3}\right) \tag{96}$$

with R_F a Carlson elliptic integral.
The second integral reads

$$I_{r,2}(\bar{r}) = \int_{\bar{r}_p}^{\bar{r}} \frac{dr}{\sqrt{R(r)}} = \frac{1}{\sqrt{1-E^2}} \int_{r_1}^{\bar{r}} \frac{dr}{\sqrt{(r_0-r)(r-r_1)(r-r_2)(r-r_3)}}. \tag{97}$$

The transformation $t = (\bar{r}-r)/(r-r_1)$ brings this integral in the form

$$I_{r,2}(\bar{r}) = \frac{2}{\sqrt{1-E^2}} \frac{\sqrt{\bar{r}-r_1}}{\sqrt{(r_0-r_1)(r_1-r_2)(r_1-r_3)}} R_F\left(\frac{r_0-\bar{r}}{r_0-r_1}, \frac{\bar{r}-r_2}{r_1-r_2}, \frac{\bar{r}-r_3}{r_1-r_3}\right). \tag{98}$$

If we substitute $\bar{r} = r_1$ in $I_{r,1}(\bar{r})$ and $\bar{r} = r_0$ in $I_{r,2}(\bar{r})$, we should get identical expressions, amounting to ΔI_r. Inspection of these expressions shows that the following equality should hold:

$$R_F(0,y,z) = \frac{1}{\sqrt{y}} \frac{1}{\sqrt{z}} R_F\left(0, \frac{1}{y}, \frac{1}{z}\right). \tag{99}$$

This can be proven by writing down the integrals, and changing one of the integral variables, say t, to $1/t$. Clearly $I_{r,2}(\bar{r}) = \Delta I_r - I_{r,1}(\bar{r})$.

We can now summarize the construction of $\bar{\vartheta}(r)$. The position on the orbit is defined by (1) the independent variable r, $\bar{r}_p \le r \le r_a$, and (2) the number of full radial half cycles m (i.e., from r_a to $\bar{r}_p$ and vice versa) that the orbit has already gone through since its 'start' at (r_0, ϑ_a). Then

$$\begin{aligned} I_r(\bar{r}) &= m\Delta I_r + I_{r,1}(\bar{r}), \qquad m \text{ even} \\ &= m\Delta I_r + I_{r,2}(\bar{r}), \qquad m \text{ odd}. \end{aligned} \tag{100}$$

To this must be added either $I_{r,1}(r_0)$ or $I_{r,2}(r_0)$, depending on the case. The angle $\bar{\vartheta}$ is determined by $n = \text{int}(I_r(\bar{r})/\Delta I_\vartheta)$ and

$$\begin{aligned} \cos\bar{\vartheta}(\bar{r}) = k\bar{v}(\bar{r}) &= k\,\text{sn}\Big(\boldsymbol{K}(k) - \frac{a}{k}\sqrt{K-E^2}\,[I_r(\bar{r}) - n\Delta I_\vartheta]\Big), \qquad n \text{ even} \\ &= k\,\text{sn}\Big(\frac{a}{k}\sqrt{K-E^2}\,[I_r(\bar{r}) - (n-1)\Delta I_\vartheta] - 3\boldsymbol{K}(k)\Big), \qquad n \text{ odd}. \end{aligned} \tag{101}$$

A little remark may be added at this point. In the travelator metric, we can construct the travelator with orbits that start off at large apocenter radii r_a, and wait for a sufficiently long time until the orbits arrive 'in the vicinity' of the black hole. From (4.5) we find in that case

$$E^2 \to 1 - \frac{2M}{r_a}, \quad K \to 1 - \frac{2M}{r_a}\sin^2\vartheta_a \quad \text{and} \quad \frac{K-E^2}{1-E^2} \to \cos^2\vartheta_a. \tag{102}$$

Hence $\Theta \to 0$, and the orbits tend toward straight lines of constant $\vartheta = \vartheta_a$ in the meridional plane.

D.2 The solution for $\overline{r}(\tau)$

We next turn to the integrals (3.10) in the expression for τ. Using integral tables [8], we find

$$\begin{aligned}-2\tau_{r,1}(\overline{r}) = -2\int_{\overline{r}}^{r_a}\frac{r^2dr}{\sqrt{R}} = (r_0+r_1+r_2+r_3)I_3'(\overline{r}) + (r_3-r_1)(r_3-r_2)I_2(\overline{r}) - \\ -\left[2r_0^2-(r_1-r_0)(r_2-r_0)\right]I_1(\overline{r}) - 2\frac{\sqrt{(r_0-\overline{r})(\overline{r}-r_1)(\overline{r}-r_2)}}{\sqrt{\overline{r}-r_3}},\end{aligned} \tag{103}$$

with

$$\begin{aligned}I_1(\overline{r}) &= 2\frac{\sqrt{r_0-\overline{r}}}{\sqrt{(r_0-r_1)(r_0-r_2)(r_0-r_3)}}R_F\left(\frac{\overline{r}-r_1}{r_0-r_1},\frac{\overline{r}-r_2}{r_0-r_2},\frac{\overline{r}-r_3}{r_0-r_3}\right)\\ I_2(\overline{r}) &= \frac{2}{3}\frac{(r_0-\overline{r})^{3/2}}{\sqrt{(r_0-r_1)(r_0-r_2)(r_0-r_3)}^{3/2}}R_D\left(\frac{\overline{r}-r_1}{r_0-r_1},\frac{\overline{r}-r_2}{r_0-r_2},\frac{\overline{r}-r_3}{r_0-r_3}\right)\\ I_3'(\overline{r}) &= \frac{2}{3}\frac{(r_0-\overline{r})^{3/2}}{\sqrt{(r_0-r_1)(r_0-r_2)(r_0-r_3)}}R_J\left(\frac{\overline{r}-r_1}{r_0-r_1},\frac{\overline{r}-r_2}{r_0-r_2},\frac{\overline{r}-r_3}{r_0-r_3},1\right).\end{aligned} \tag{104}$$

After m radial half cycles we obtain

$$\begin{aligned}\tau_r(\overline{r}) &= m\tau_r(\overline{r}_p) + \tau_{r,1}(\overline{r}), \qquad m \text{ even}\\ &= (m+1)\tau_r(\overline{r}_p) - \tau_r(\overline{r}) \equiv m\tau_r(\overline{r}_p) + \tau_{r,2}(\overline{r}), \qquad m \text{ odd}.\end{aligned} \tag{105}$$

To this must be added either $\tau_{r,1}(r_0)$ or $\tau_{r,2}(r_0)$, depending on the case. The integral in ϑ (reads)

$$\tau_\vartheta(\overline{\vartheta}) = a^2\int_{\vartheta_a}^{\overline{\vartheta}}\frac{\cos^2\vartheta d\vartheta}{\sqrt{\Theta(\vartheta)}} = \frac{a}{\sqrt{K-E^2}}k^3\int_{\overline{\nu}}^{1}\frac{\nu^2d\nu}{\sqrt{1-\nu^2}\sqrt{1-k^2\nu^2}}. \tag{106}$$

When $\vartheta_a \leq \overline{\vartheta} \leq \pi - \vartheta_a$ we obtain

$$\tau_{\vartheta,1}(\overline{\nu}) = \frac{ak}{\sqrt{K-E^2}}\left[\boldsymbol{K}(k) - \boldsymbol{E}(k) - F(\arcsin\nu,k) + E(\arcsin\nu,k)\right], \tag{107}$$

or

$$\tau_{\vartheta,1}(\overline{\nu}) = \frac{a}{\sqrt{K-E^2}}\frac{k^3}{3}\left[R_D(0,1-k^2,1) - \overline{\nu}^3R_D(1-\overline{\nu}^2,1-k^2\overline{\nu}^2,1)\right], \tag{108}$$

while for the second branch, for decreasing $\overline{\vartheta}$, we obtain

$$\tau_{\vartheta,2}(\overline{\nu}) = \frac{ak}{\sqrt{K-E^2}}\left[3[\boldsymbol{K}(k) - \boldsymbol{E}(k)] + F(\arcsin\nu,k) - E(\arcsin\nu,k)\right], \tag{109}$$

or

$$\tau_{\vartheta,2}(\overline{\nu}) = \frac{a}{\sqrt{K - E^2}} k^3 \left[R_D\left(0, 1 - k^2, 1\right) + \frac{\overline{\nu}^3}{3} R_D\left(1 - \overline{\nu}^2, 1 - k^2\overline{\nu}^2, 1\right) \right]. \tag{110}$$

We recall that the angle $\overline{\vartheta}$ is determined by (D18), and thus

$$\begin{aligned} \tau_\vartheta(\overline{\nu}) &= n\tau_{\vartheta,1}(-1) + \tau_{\vartheta,1}(\overline{\nu}), \qquad n \text{ even} \\ \tau_\vartheta(\overline{\nu}) &= (n-1)\tau_{\vartheta,1}(-1) + \tau_{\vartheta,2}(\overline{\nu}), \qquad n \text{ odd}. \end{aligned} \tag{111}$$

Finally,

$$c\left[\tau\left(\overline{r}, \overline{\vartheta}\right) - \tau_0\right] = \tau_r(\overline{r}) + \tau_\vartheta(\overline{\nu}). \tag{112}$$

Since $\overline{\nu}(\overline{r})$ according to (D18), this expression can be inverted to yield $\overline{r}(\tau)$.

Author details

Herwig Dejonghe
Sterrenkundig Observatorium, Ghent University, Gent, Belgium

*Address all correspondence to: herwig.dejonghe@ugent.be

References

[1] Dejonghe H. Spherical Cosmological Models: An Alternative Cosmology. 2018. arXiv:1808.10278v3

[2] Hackmann E, Lämmerzahl C, Kagramanova V, Kunz J. Analytical solution of the geodesic equation in Kerr-(anti) de Sitter space-times. Physical D Review. 2010;**81**:044020

[3] Chandrasekhar S. The Mathematical Theory of Black Holes. New York: Oxford University Press; 1992

[4] Landau LD, Lifshitz EM. The Classical Theory of Fields. Oxford: Pergamon Press; 1971

[5] Khatsymovski VM. On the Kerr metric in a synchronous reference frame. International Journal of Modern Physics D: Gravitation; Astrophysics and Cosmology. 2021;**30**:2150071

[6] Adler R, Bazin M, Schiffer M. Introduction to General Relativity. New York: McGraw-Hill; 1975

[7] Doneva D, Yazadjev S. Nonradial oscillations of anisotropic neutron stars in the Cowling approximation. Physical Review D. 2012;**85**:124023

[8] Carlson BC. A table of elliptic integrals of the third kind. Mathematics of Computation. 1988;**183**:267

Chapter 6

Closed Loop Universe and the Topology of Totality

Michael G. Albrow

Abstract

The author suggests a cosmology based on just two principles: (1) no boundaries to spacetime and (2) no physical infinities. On the hypothesis that these are exactly true, and that charge conjugation, parity and time reversal invariance (CPT) is a perfect symmetry at the big bang, answers are provided to many puzzles, such as what happened to the antimatter, what happened before the big bang, and what will happen in the far future? Consequences are that our universe U is closed (with positive curvature) and so the total energy, angular momentum and charge of the universe must be zero. At the big bang a universe-antiuniverse pair, $U\overline{U}$ is created as a virtual fluctuation in what we may call Outer SpaceTime, along with a vast number of other disconnected universes—the multiverse. The CPT-conjugate $\overline{U}$ has identical physics (and matter content) to U; other $U\overline{U}$ pairs in the multiverse have different physics parameters. A consequence is that *time* at the fundamental particle level *à la* Feynman must be distinct from thermodynamic time. The $U\overline{U}$ forms a closed loop.

Keywords: cosmology, loop universe, baryon asymmetry, CPT, multiverse, time

1. Introduction

If charge conjugation, parity and time reversal invariance (CPT) is a perfect symmetry applied to the universe, the question has long been asked, and is still considered a major puzzle: "Where is the antimatter?" or now more usually: "What happened to the antimatter?" as we assume that the origin was perfectly symmetric.

Already in 1956, shortly after the discovery of the antiproton, Goldhaber published [1] a "speculation" to preserve particle—antiparticle symmetry that " ... the cosmos and its possible counterpart, the "anticosmos," are somehow separated from the very beginning of their existence."

In 1972, Omnes discussed the problem [2] and argued that there should exist as many antigalaxies as galaxies. No credible matter–antimatter separation mechanism was proposed and he listed many problems, for example, the consequences of galaxy-antigalaxy collisions.

Later that year Davies proposed a model [3] in which time reverses between successive cycles of an oscillating universe. In 1973, the author proposed [4] that CPT conservation at the bounces leads to a different interpretation, that our universe U and an antiuniverse $\overline{U}$ came into existence together, and will eventually come back together in a common "big crunch." A baryon world-line is a closed loop, and the author noted that this is "reminiscent of fluctuations of the vacuum in quantum electrodynamics, in which a particle-antiparticle pair appear in space and then annihilate." This $U\overline{U}$ pair has total charge, baryon number and lepton number all equal to zero, but the matter is all in U and the antimatter all in $\overline{U}$. There is no need for some "Great Annihilation" of antimatter after the big bang. The total energy also has to be zero, because space is closed, and so electrical and gravitational fields must be confined. (That is not of course the case for a virtual e^+e^- pair, as proven by the Lamb shift, etc.) This suggestion answered the question: "*What happened to the antimatter?*" It is in the parallel $\overline{U}$, disconnected from U except at the big bang and a future big crunch. (Related work in the 1960's by Sakharov [5] had been published only in Russian.) This implies an "Outer SpaceTime" in which such fluctuations occur.

Tryon then published a note [6] in which he also suggested that the universe might be a vacuum fluctuation with total energy $E(U) = 0$. His model differed from Ref. [4] in that our universe should consist equally of matter and antimatter, as previously discussed by Omnes and others.

There has been enormous progress in cosmology in the intervening 50 years, and we can now state with confidence that all of the universe presently accessible to observation grew out of a big bang some 13.8×10^9 years ago, that it has been expanding ever since, and that the expansion appears to be accelerating.

2. Two hypothetical principles

If one supposes two simple principles to be absolutely true, a particular topology of the universe follows naturally. These principles are:

P1: Spacetime has no boundaries. Einstein's idea that space may finite but has no boundaries is familiar [7], but I extend this to spacetime.

P2: No physical infinities. The realm of infinities should be confined to mathematics. While 10^{183} is a very big number it is not "almost infinity."

Strictly applying P1 and P2 has many radical consequences. Singularities (with infinite density) are forbidden. While these principles are not obviously falsifiable by experiment, they lead to a cosmology that provides answers to some of the mysteries: the matter dominance of our universe, what happened before the big bang, and what is its ultimate fate in "deep time"? The proposed model can be called Closed Loop Universe (CLU). *This will require making a distinction between the "thermodynamic time" of everyday experience and the "world-line time" of elementary particles.*

3. P1: Spacetime has no boundaries

The concept that the three dimensions of space can form a finite volume without boundaries, analogous to the two dimensions of the surface of a sphere, was proposed by Einstein [7], the mathematical form having been invented in 1854 by Riemann. Being unable to formulate boundary conditions for "spatial infinity," Einstein said: "If it were possible to regard the universe as a continuum which is finite (closed) with respect to its spatial dimensions, we should have no need at all of any such boundary conditions." The total volume of a closed spherical 3-space S^3 of radius R is $V = 2\pi^2 R^3$.

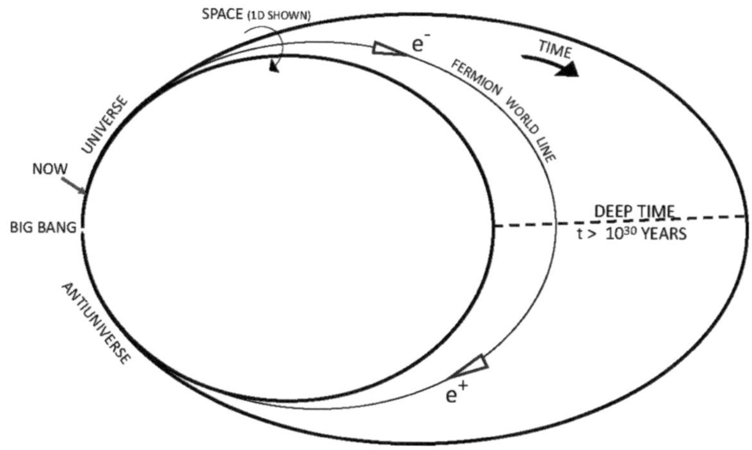

Figure 1.
Sketch of the toroidal topology of the CLU version of the universe, with world-line time as a closed loop. The big bang marks the origin of our universe and the symmetrical antiuniverse. Only one of the three space dimensions is shown. Thermodynamic time is different from world-line time and increases away from the big bang in both universe and antiuniverse.

We extend the notion of closed and finite 3-space to 4-dimensional spacetime, *à la* Minkowski. Let $x_1 = x$, $x_2 = y$, $x_3 = z$, and $x_4 = ict$. While the three spatial coordinates have arbitrary orientation but are orthogonal to each other, time is orthogonal to the dimensions of space in the same sense that imaginary and real coordinates are orthogonal in the complex plane. We suppose that not only do x_1, x_2, x_3 form a closed spherical space S^3, but x_1, x_2, x_3, and x_4 form a closed (Euclidean) spacetime S^3T^1. In x,y,z,t coordinates this is a toroid.

Consider a tube of radius r and length L and join the ends to form a torus. On its 2-dimensional outer surface, the circumference (length 2πr) represents one space dimension and the orthogonal coordinate along L represents time, t, a topology S^1T^1. Replacing the one space dimension with three gives a topology S^3T^1, in which time, like space, is finite but unbounded. The three space dimensions form the 3-dimensional surface of a 4-dimensional hypersphere. Without changing the topology, make r(t) a function of time by constricting the tube at one time to be very small (at the big bang) but not zero (no infinite curvature of spacetime, no singularities!). The concept of the CLU is shown in **Figure 1**, with only one space dimension shown; it is not just closed space but *closed spacetime*.

The universe is expanding with the radius r(t) apparently increasing faster than linearly at the present time, attributed to a cosmological constant or dark energy; if true this must be temporary as "forever" is not allowed. Consider the world-line of each lepton or quark—in a Feynman diagram of the whole $U\overline{U}$ pair! Fermion world-lines have zigzags corresponding to pair creations and annihilations.[1]

In Ref. [4], the author had suggested a future collapse to a big crunch, which was identified with the big bang. The problems of entropy at that event, together with the hypothesis of closed loop time, can be avoided by instead identifying the far future of our universe with the far future (deep time) of the "parallel" antiuniverse.

[1] For simplicity we can ignore the weak interaction and neutrinos, but they do not affect the argument. We just say "fermion world-lines" which have no beginning and no end but are finite.

This CLU is a "closed timelike loop," but unlike closed timelike loops that have been much discussed and found to raise major issues in causality, this does not. It is not possible to travel back in time. Hawking's "Chronology Protection Conjecture" [8] applies.

This does not mean that any event is *repeated*; world-line time is cyclic but there is just one cycle, and *that is it*. It is easy to see how, in this topology, the questions: "What happened before the big bang?" and "Does time have an end?" have no more meaning than "Where is the edge of the Earth's surface?" The future of our universe is finite, as is that of the parallel antiuniverse, and in deep time they will become one.

4. P2: No physical infinities

Infinities abound in mathematics, but the author proposes that they do not occur in the physical universe. If this hypothesis P2 is true, there cannot be an infinite number of particles (or universes). The total volume of space in the universe must be finite. To say that the volume is finite, we must define a measure. The only natural candidate is the Planck volume $V_{Planck} \sim L^3_{Planck} \sim 4 \times 10^{-105}$ m^3, where $L_{Planck} = \sqrt{\hbar G_N / c^3}$. Today the volume of the observable universe is $\sim 10^{78}$ m^3 $\sim 10^{183} \times V_{Planck}$. The number is increasing as the Universe expands, but cannot become infinite. The accelerated expansion attributed to dark energy must be temporary. With no infinite quantities, we should have no infinite densities or singularities, neither at the big bang nor at the center of black holes.

If fermion world-lines are continuous through the big bang, there are $\gtrsim 10^{80}$ fermion lines threading the constriction.[2] *If* each fermion has to occupy a distinct Planck volume (which may not be true), they could fit in a volume with $R_{\rm neck} \approx 10^{-8}$ m.

It has been suggested by Pegg [9] that a very few fermions (or even only one of each type) could thread the big bang some 10^{80} times. In the extreme limit, every electron could actually be the same one. When Feynman drew world lines as zigzags in spacetime, he noted (probably very tongue-in-cheek) this possibility within our universe, but to be true it needed an equal amount of antimatter and matter, and therefore required an unlikely large-scale separation mechanism. It would have the advantage of explaining why all electrons (and up quarks, etc.) have the same mass. They are the same particle!

In 1967, Sakharov [5] gave three conditions that should be satisfied to account for the matter–antimatter asymmetry in our universe *following* a big bang: CP-violation, baryon number violation and non-equilibrium dynamics. While non-equilibrium dynamics no doubt prevailed at the big bang, his first two principles are circumvented by considering not just "our universe" but the $U\overline{U}$ pair. The CP-violation observed in weak interactions may have little or nothing to do with baryon number violation.

Models in which the big bang is preceded by a "big crunch" with a bounce have to address the Second Law of Thermodynamics. What happens to the entropy? This difficulty is avoided with this radical proposal of a distinction between (a) "world-line time" in Feynman diagrams and with his idea [10] that positrons can be considered as electrons moving backwards in time, and (b) "thermodynamic time" with entropy increasing, teacups breaking and our memories forming. The author will comment on

[2] Some antimatter world lines may have emerged from the big bang along with all the matter lines, but not necessarily.

thermodynamic time in Section 6, but for the rest of this chapter by "time" the author means world-line time unless otherwise stated.

This closed space, really closed spacetime, is different from that of a black hole, which has mass-energy and can have electric charge, and so has external electromagnetic and gravitational fields.

The author will say nothing about any possible small extra dimensions except that they should obviously all be closed.

5. The big bang and inflation

A motivation for inflation [11] was the so-called "flatness problem" that the overall density of the universe ρ seems to be equal to the critical density for closure ρ_{crit}. Possibly inflation (but not *eternal inflation*) can be incorporated into the CLU model. In 1982, Vilenkin [12, 13] also proposed that the universe is created by quantum tunneling from nothing, again noting an analogy with e^+e^- pair production. He said: "The only verifiable (in principle) prediction of the model is that the universe must be closed. However Guth has argued [14] that the inflationary universe almost certainly overshoots, so that $\rho = \rho_{crit}$ with a very high accuracy even at the present time. This means that we shall have to wait for a *long time* (the author's italics) until the sign of $(\rho - \rho_{crit})$ can be determined experimentally." In the CLU model, the density parameter ρ exceeds ρ_{crit} but by a tiny amount that will never be measured if inflation is correct. Guth's *long time* should be greater than the proton lifetime, for reasons discussed in the next section.

The present data is consistent with $\rho/\rho_{crit} = 1$ but with about 1% uncertainty, that is, visible 3-space is consistent with being flat overall. Local regions of curvature are caused by matter, but in inflationary scenarios the visible universe is a tiny fraction of the whole. The closed 2D surface of the Earth has regions that are locally flat and other regions with positive curvature (hills) and negative curvature (saddles). If the 3-space of the whole universe is perfectly flat (ignoring density fluctuations such as galaxies!) or has negative curvature, it must be infinite, which violates P2 and is incompatible with this model.

According to CLU at the big bang there cannot be a "singularity"; the volume is not zero but the total energy is zero, with positive matter energy canceled by negative gravitational potential energy. In inflationary theories the total energy of the universe can be zero [11, 15, 16].[3]

6. Deep time: The far future

Since in the CLU fermion world-lines are endless and continuous through the common big bang the total number of leptons and quarks in U must be the same as the number of antileptons and antiquarks in $\overline{U}$. The physics in the "parallel" antiuniverse is the same as in our universe. Both U and $\overline{U}$ are closed 3-spaces, with no connection except at the common big bang and the common far future, *deep time*.

---

[3] One may ask whether zero energy is incompatible with the Heisenberg Uncertainty Relation $\Delta E \geq \hbar/\Delta t$, so absolutely zero energy implies an infinite time, which is forbidden in CLU. However, the duration of the CLU should be greater than 10^{30} years, so perhaps the point is moot.

How can U and $\overline{U}$ be connected in the far future? While the physics is identical in U and $\overline{U}$ the evolution will be very different. Galaxies and stars and planets of antimatter (according to us!) may form, and even life, if inevitable, will begin and evolve with as much variety as life on Earth and elsewhere in our universe. If protons are absolutely stable, planets and anti-planets would likely persist to deep time (black holes will presumably not swallow them all due to the expansion) and such a matching would be impossible.

That impossible matching scenario can be avoided if there are no complex objects, such as teacups, molecules or even protons surviving, but only stable elementary particles. Supposing the complete list to be: $e^+, e^-, \nu_i, \overline{\nu}_i$, photons and maybe gravitons. In that case, a matching of U and $\overline{U}$ seems to require proton decay, $p \rightarrow e^+ +$ photons (directly or *via* light mesons). This violates separate baryon and lepton number conservations, but those "laws" are not fundamental; they were invented to account for the stability of matter. But fermion conservation still applies; fermion world-lines have no ends. But in the absence of any interactions between them, they should not be considered as localized, but described by their probability waves $\psi(x, y, z, t)$. Quantum mechanics reigns! In this very cold far future, the wavelengths are vast, even electrons and positrons (and neutrinos) will have probability waves extending over parsecs! Locality (and matching of fermions and antifermions) would be less of an issue. Entropy will have become irrelevant.

Hawking [17] once proposed that "time reverses" when the universe has reached its maximum expansion. He said: "What would happen if and when the universe stopped expanding and began to contract? Would the thermodynamic arrow reverse and disorder begin to decrease with time? ... broken cups gathering themselves together off the floor ... ?" Later he wrote [18] that he was convinced that this was his "greatest mistake in physics: the universe would not return to a smooth state in the collapse. This would mean that the arrow of time would not reverse." By making a distinction between thermodynamic time and world-line time, that statement can be true of thermodynamic time.

Steinhardt and Turok [19–21] proposed a cyclic model of a universe that is infinite and flat. An "accelerated expansion, caused by dark energy, is necessary to dilute the entropy, black holes and other debris." They said in deep time "the universe is returned to its original pristine vacuum state before it begins to contract (due to a negative potential energy), bounce and begin a new cycle." In the CLU model, the entropy is diluted, even becomes meaningless, allowing the U and $\overline{U}$ to merge and complete the closed time loop—not only world-line time but thermodynamic time.

7. Some other related cosmologies

In recent years, several authors have developed in detail the concept of universe-antiuniverse pair creation. These have similarities with the CLU model but generally involve infinite time.

In 1974, Gott [22] also considered a time-symmetric cosmology, treating the three Minkowskian regions of spacetime at an initial singularity (absolute past, absolute future and "elsewhere") as if they were distinct universes. The absolute past is C-, P- and T-inverted (antimatter). This differs from the CLU in that his time is infinite in both directions (absolute past and future), and his "elsewhere" is a tachyonic universe.

Hawking has also discussed [17] a finite spacetime with no boundaries in the context of a quantum theory of gravity. He said: "One could say: "The boundary condition of the universe is that it has no boundary." The universe would be completely self-contained and not affected by anything outside itself. He went on to say: "I'd like to emphasize that this idea that time and space should be finite without boundary is just a *proposal*: it cannot be deduced from some other principle. Like any other scientific theory, it may initially be put forward for aesthetic or metaphysical reasons, but the real test is whether it makes predictions that agree with observations."

The requirement that the total energy-momentum, angular four-momentum and electric charge of a closed space are zero restricts the classes of allowed universes (solutions to the equations of general relativity) that can arise from vacuum fluctuations. Ferrando, Lapiedra and Morales [23] showed that closed and flat FRW universes can satisfy these conditions irrespective of conformal transformations, including dilatations, rotations and translations on all spacelike 3-surfaces. However, this is not the case for all universes, e.g., open FLRW universes or the Bianchi V family of universes [24]. It would be interesting to analyze in this way other metrics, such as the Lemaître-Tolman-Bondi universe, which may explain the accelerated expansion without the need for dark energy [25], Stephani universes [26], etc.

Boyle, Finn and Turok [27] applied CPT as an exact symmetry at the big bang, such that the *thermodynamic* arrow of time (defined by entropy increasing) reverses.

In 1994, Rosen [15] and Cooperstock and Israelit [16, 28] considered the total energy-momentum of open and closed universes in general relativity, in the absence of a cosmological constant, and found $E = 0$ for any homogeneous isotropic universe described by a FLRW metric. Recently, several authors [29–31] have supported the idea that the net energy of the universe must be equal to zero.

Robles-Pérez has also recently developed [32, 33] ideas about the quantum creation of universe-antiuniverse pairs.

The CLU is incompatible with an ultimate "big rip" [34], a future singularity suggested in an eternally accelerating universe.

8. Black holes

The "no hair" theorem says that black holes may have electric charge, four-momentum and angular momentum but no other *external* attributes. An outside observer cannot say anything about the baryon number of a black hole, as there is no long range force that is different for baryons and antibaryons; matter and antimatter black holes look identical. A closed universe is not a black hole; there is nothing external. Hawking proposed that black holes have an associated temperature given by $T_{BH} = \hbar c^3/8\pi GMk_b$, and will evaporate if in a heat bath at a lower temperature. The thermal radiation emitted is equally matter and antimatter, and if that were the case with complete evaporation a black hole formed of matter would violate baryon number conservation on a large scale. However, this conclusion is perhaps not inescapable; indeed, Hawking himself said [17] "What happens when the mass of the black hole becomes extremely small is not quite clear, but *the most reasonable guess* (my italics) is that it would disappear completely … " and "total evaporation is just an assumption." Arguments about the "inevitable singularity" at the center of a black hole rest on untested assumptions about gravitation and other possible interactions at the Planck scale. As fermion conservation would forbid complete evaporation of matter, stable cores would remain, perhaps very small.

9. Entropy considerations

The concept of an oscillating universe with multiple (even an infinite number of) bounces, but without closed loop time, is old [35]. A serious issue [36] has always been the sacrosanct Second Law of Thermodynamics, stating that the entropy of a closed system, if not in equilibrium, must always increase. In models where the universe reverses its expansion in the far future and collapses down to a "big crunch" which is identified with a future big bang (or one in our past), we would be forced to abandon the Second Law. Also, in Ref. [4] there is no symmetry between the early histories of U and $\overline{U}$.

The CLU model avoids those issues. The U and $\overline{U}$ are matched in the far future. This requires making a distinction between "world-line time" and thermodynamic time. In Feynman diagrams of particle interactions, electrons are shown as moving forward in time, and positrons are shown as moving backward [10]. This is purely conventional and no fundamental asymmetry is implied that would violate CPT. That *time* at the level of elementary particles can be called "world-line time." A Feynman diagram of the whole CLU would show no distinction between the upper (U) and lower ($\overline{U}$) branches in **Figure 1**. The time in the arena of complex objects, from molecules to breaking teacups to brains (and hence consciousness), can be called thermodynamic time and flows from left to right in **Figure 1** in both U and $\overline{U}$. But this must (in the CLU model) become irrelevant in the far future common to both U and $\overline{U}$.

10. Experimental tests

While the principles of no physical infinities and no boundaries to spacetime could only be disproved by observing such infinities or boundaries, assuming them to be true imposes constraints on cosmology that can in principle be tested. Closed space requires the cosmological density parameter $\rho/\rho_{crit} > 1.0$, consistent with observations; if it is found to be definitely less than 1.0 it is likely that P1 and P2 are wrong. The assumption of baryon and lepton number conservation is being actively tested by many experiments; we note that they may theoretically be violated [37] even in the Standard Model. But the supposition of endless fermion world-lines and that singularities are forbidden predicts that black holes cannot evaporate to nothing; microscopic remnants carrying the full baryon number and lepton number remain. They will have little or no electric charge and would be largely invisible except through their (very strong at small distances) gravitational fields. If they exist such cores might be detectable and would be dark matter candidates.

Even though it may be practically impossible to either disprove or prove the CLU model, its strength is that it can potentially explain many current mysteries. In this respect, it is similar to other ideas about the multiverse and is based on only P1 and P2 with CPT conservation.

11. Outer SpaceTime: Totality

This section is quite different in that I speculate about matters outside our closed spacetime and therefore in principle *unknowable*; that need not prevent us from thinking about it. If our $U\overline{U}$ is a fluctuation, there should be many other $u\overline{u}$ pairs in some "Outer SpaceTime." What is the topology of Outer SpaceTime? It can also be finite with no boundaries, as in **Figure 2**. Perhaps the finite closed spacetime I have presented is *everything*, and that is it. If so it seems highly improbable that the laws and constants (forces, masses of particles, etc.) of physics are just right for life (the Goldilocks Principle applied not to exoplanets but to the universe). This is the basis for the many ideas about a multiverse.

As Einstein realized for our universe, a matter-dominated universe without a cosmological constant is unstable and must expand or contract. That will not be the case for Outer SpaceTime, because they $u\overline{u}$ have zero total energy-momentum and there may be no interaction between them. Therefore, the "radius" of Outer Space can be constant all around the Time-loop. Remarkably, this picture of Outer Space is something like the old "Steady State" model of Bondi, Gold and Hoyle [38, 39], being (on average) the same at all Times and Places. This is the Perfect Cosmological Principle, that the Universe should look the same at all times and places (apart from local fluctuations) and it is satisfied by the CLU model, but only in Outer SpaceTime. It achieves this without having to invent a mechanism to continuously create protons from an expanding vacuum, without having an infinitely long time duration, and without having to explain such facts as the cosmic microwave background and primordial light element formation.

Outer Space can even be 1-Dimensional, a Line. Join the ends to make a Ring, and propagate through Time to make a Tube, then join the ends to make a Torus; $u\overline{u}$ fluctuations on the Line are on the surface of that Torus. There may be 2, 3, or any number of Outer Space dimensions (but they are harder to draw!).

In the CLU with the universe and antiuniverse coming into existence together at t_0 and matching in the far future at t_{max}, in a sense U and $\overline{U}$ exist as parallel but disconnected universes at the present (psychological) time t (i.e., now). Returning to our analogy with a virtual e^+e^- world-line loop in our spacetime, arising as a vacuum fluctuation thanks to the uncertainty principle $\Delta E > h/\Delta t$. The analogy is that $u\overline{u}$ arise as virtual fluctuations but in some Outer SpaceTime, a different arena. Naturally, there would be a vast number of such fluctuations, that is, other $u\overline{u}$ pairs; a version of the multiverse idea, shown in **Figure 2** as a cartoon. Having no connections between

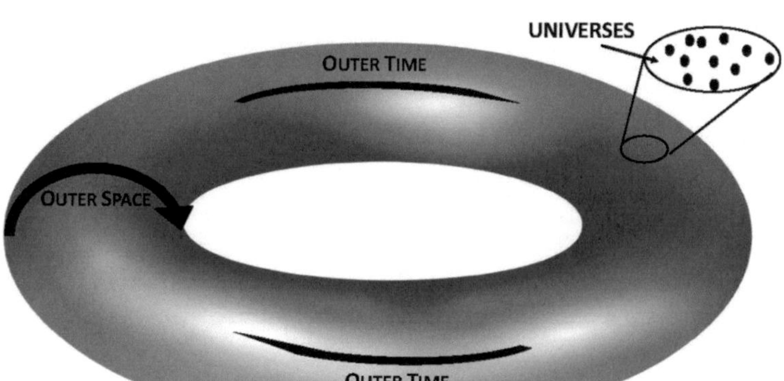

Figure 2.
Closed Loop model of Outer SpaceTime, with just one Outer Space dimension shown. A vast number of $u\overline{u}$ pairs appear on the surface as virtual fluctuations. The directions of the arrows is arbitrary (or meaningless!).

them all, physical parameters can be different; ours just happening to be one where they are just right for life.

To quote Vilenkin [12], who gave a mathematical description of the process of universe creation by tunneling of a de Sitter-Hawking-Moss *instanton* from "nothing" ($\equiv$ Outer Space): "*Nothing* is the realm of unrestrained quantum gravity; it is a rather bizarre state in which all our basic notions of space, time, energy, entropy etc., lose their meaning."

The Perfect Cosmological Principle is also satisfied by Linde's theory of eternal chaotic inflation [40] and by Aguirre and Gratton's cosmology [41, 42]. One key difference is that these multiverses are "eternal" while here we forbid eternity. The direction of Time around the torus can be reversed and nothing changes (there is no entropy in Outer SpaceTime). It is *possible* that the $U\overline{U}$ could be connected by a network of "filaments," perhaps even with a fractal pattern such as proposed by Linde. In the CLU such connections are unnecessary and unwelcome, as they destroy the independence of the universes necessary for the Goldilocks Principle to apply in our universe and allow life.

12. Conclusions

The Closed Loop Universe model presented here is based on the two principles, assumed exact, of no boundaries to spacetime and no physical infinities. The CLU model answers the question "What happened to the antimatter?" It is all there in a distinct closed space, a parallel universe disconnected from ours except at the common big bang and in the future at $t \gtrsim 10^{32}$ (years).

The CLU relates to ideas of Einstein, Feynman, Hawking, Steinhardt and Hoyle among many others. The topology of our universe is a S^3T^1 toroid with world-line time (*à la* Feynman) as a closed loop. Thermodynamic, entropic or experienced time are not the same as world-line time. The universe and antiuniverse came into existence together, analogous to a virtual fermion loop but on a cosmic scale. Among the many radical consequences are statements about what happened before the big bang and what will happen in the far future, the accelerated expansion (attributed to dark energy) must be transitory, the total energy, angular momentum and charge of the universe must be zero, and the cosmological density parameter $\rho/\rho_{crit} = 1.0 + \varepsilon$ with ε positive but unmeasurably small. Baryon number and lepton number are not separately conserved, protons must decay to positrons + photons, but fermion world-lines have no ends. Black holes cannot evaporate completely, and the cosmic photon:baryon ratio can be explained without a "Great Annihilation."

Finally, and very speculatively, the author described a closed and finite toroidal Outer SpaceTime in which the CLU is a zero-energy fluctuation, along with a huge number of others (the multiverse) with different physical parameters. Outer SpaceTime need not expand or contract; if static it is a revised version of the "steady state model" obeying the Perfect Cosmological Principle of Hoyle.

Acknowledgements

The author acknowledges valuable discussions with A.J. Stebbins III, and the Fermi National Accelerator Laboratory for support.

Author details

Michael G. Albrow
Fermi National Accelerator Laboratory, Batavia, USA

*Address all correspondence to: albrow@fnal.gov

References

[1] Goldhaber M. Speculations on cosmogony. Science. 1956;**124**:218

[2] Omnes R. Matter-antimatter separation and the antimatter problem in cosmology. In: Symposium on the Physics of Dense Matter. Boulder, CO; 1972

[3] Davies PCW. Closed time as an explanation of the black body background radiation. Nature Physical Science. 1972;**240**:3

[4] Albrow MG. CPT conservation in the oscillating model of the universe. Nature Physical Science. 1973;**241**:56

[5] Sakharov AD. Violation of CP invariance, C asymmetry, and baryon asymmetry of the universe. Pis'ma v Zhurnal Èksperimental'noi i Teoreticheskoi Fiziki. 1967;**5**:32

[6] Tryon EP. Is the universe a vacuum fluctuation? Nature. 1973;**246**:396

[7] Einstein A. Cosmological Considerations on the General Theory of Relativity. Wissenschaften: Sitzungsberichte der Preussischen Akad; 1917

[8] Hawking SW. The chronology protection conjecture. Physical Review. 1992;**D46**:603

[9] Pegg DT. Absorber theory of radiation. Report Progress in Physics. 1975;**38**:1339

[10] Feynman RP. The theory of positrons. Physics Review. 1949; **76**:749

[11] Guth A. Eternal inflation. Annals of New York Academy of Sciences. 2001; **950**:66

[12] Vilenkin A. Birth of inflationary universes. Physical Review D. 1983;**27**: 2848

[13] Vilenkin A. Creation of universes from nothing. Physics Letters. 1982; **117B**:25

[14] Guth AH. MIT preprint CTP-976. 1982

[15] Rosen N. The energy of the universe. General Relational Gravitational. 1994; **26**:319

[16] Cooperstock FI. Perspectives on the energy of the universe. General Relational Gravitational. 1994;**26**:323

[17] Hawking S. A Brief History of Time. Bantam Press, Transworld Publishers Ltd.; 1988. pp. 149-150

[18] S. Hawking in Hawking and Penrose, The Nature of Space and Time, Princeton University Press. 1996. p. 101

[19] Steinhardt PJ, Turok N. A cyclic model of the universe. Science. 2002; **296**:1436

[20] Steinhardt PJ, Turok N. Cosmic evolution in a cyclic universe. Physical Review D. 2002;**65**:126003

[21] McFadden PL, Turok N, Steinhardt P. Solution of a Braneworld big crunch/big bang cosmology. Physical Review D. 2007;**76**:104038

[22] Gott JR III. A time-symmetric matter, antimatter, tachyon cosmology. The Astrophysical Journal. 1974;**187**:1

[23] Ferrando JJ, Lapiedra R, Morales JA. Creatable universes. Physical Review D. 2007;**75**:124003

[24] Estabrook FB, Wahlquist HD, Behr CG. Dyadic analysis of spatially

homogeneous world models. Journal of Mathematical Physics. 1968;**9**:497

[25] Alnes H, Amarzguioui M, Grøn Ø. Inhomogeneous alternative to dark energy? Physical Review D. 2006;**73**: 083519

[26] Stephani H et al. Exact Solutions of Einstein's Field Equations. Cambridge University Press; 2003

[27] Boyle L, Finn K, Turok N. The big bang, CPT, and neutrino dark matter. Physical Review Letters. 2018;**121**:251301

[28] Cooperstock FI, Israelit M. The energy of the universe. Foundations of Physics. 1995;**25**:631-635

[29] Salti M. On the energy-momentum of closed universes. Czechoslovak Journal of Physics. 2006;**56**:177

[30] Aygun S, Baysal H, Tarhan I. Energy momentum of Marder universe in teleparallel gravity. Intenational Journal of Modelling Physics. 2007;**46**:2607

[31] Vargas T. The energy of the universe in teleparallel gravity. General Relational Gravitation. 2004;**36**:1255

[32] Robles-Pérez SJ. Quantum creation of a universe-antiuniverse pair. Acta Physica Polonica B, Proceedings Supplement. 2020;**13**:325

[33] Robles-Pérez SJ. Hartle-Hawking's vacuum is full of Vilenkin's universe-antiuniverse pairs. e-Print arXiv: 2110.06521 [gr-qc]. 2021

[34] Caldwell RR, Kamionowski M, Weinberg NN. Phantom energy and cosmic doomsday. Physical Review Letters. 2003;**91**:071301

[35] Tolman R. Relativity, thermodynamics and cosmology: Oxford Clarendon Press; 1934. ISBN 0-486-65383-8

[36] Tolman RC. On the problem of entropy in the universe as a whole. Physics Review. 1931;**37**:1639

[37] t' Hooft G. Symmetry breaking through Bell-Jackiw anomalies. Physical Review Letters. 1976;**37**:8

[38] Bondi H, Gold T. The steady-state theory of the expanding universe. Monthly Notices of the Royal Astronomical Society. 1948;**108**:252-270

[39] Hoyle F. A new model for the expanding universe. Monthly Notices of the Royal Astronomical Society. 1948; **108**:372-382

[40] Linde A. Inflationary cosmology. Lecture Notes in Physics. 2008;**738**:1-54. e-Print: 0705.0164 and references therein

[41] Aguirre A, Gratton S. Steady-state eternal inflation. Physical Review D. 2002;**65**:083507

[42] Aguirre A, Gratton S. Inflation without a beginning, a null boundary proposal. Physical Review D. 2003;**67**: 083515

Chapter 7

Non-Particulate Quantum States of the Electromagnetic Field in Expanding Space-Time

Philip Broadbridge, Sarah Becirevic and David Hoxley

Abstract

A quantum field has been coupled to a space-time with accelerating expansion. Dynamical modes are destabilised successively at shorter material wavelengths as they metamorphose from oscillators to repellers. Due to degeneracy of energy levels, the number of unstable modes increases at an accelerating rate, sufficient to account for a significant proportion of cosmic energy. For the subsystem spanned by a finite basis of unstable runaway modes, the quantum Hamiltonian is unbounded below. There is no Bogoliubov transformation by which that subsystem Hamiltonian can be expressed as a linear combination of number operators. For the remaining subsystem spanned by an infinite number of oscillator modes, there is an appropriate vacuum state in a Fock-Cook representation of the field algebra. The massless quantum vector field of electromagnetism is considered when it is minimally or more generally coupled to an expanding space-time. For a significant class of models, including minimal coupling models and the exponential de Sitter universe coupled to the Ricci curvature tensor, the field equations are equivalent to the Proca equation with time-dependent mass.

Keywords: vacuum instability, electromagnetic field, FLRW, number operators, dark energy

1. Introduction

The simplest Bosons that we encounter are phonons that mediate the equally spaced energy levels of a quantised harmonic oscillator. In general, small oscillations of a many-body system about a stable equilibrium state can be approximated as a Hamiltonian system with positive definite quadratic Hamiltonian. It is well known that for such a system, there exists a linear canonical (or symplectic) transformation on phase space by which the Hamiltonian is seen to be that of a system of independent harmonic oscillators, with angular frequencies $\omega_j; j = 1 \cdots N$, that are the frequencies of the elementary excitations of the system with N degrees of freedom. That concept can be extended to infinite degrees of freedom in the guise of field theories. Relativistic free field theories are generally based on invariant linear wave equations that involve well defined particle mass and spin that are invariants of the inhomogeneous Lorentz group. They can still be represented mathematically as a system of independent harmonic oscillators. In the early years of quantum field theory, that approach was developed systematically for the electromagnetic field as well as for a massive scalar field [1]. Each component oscillator can be quantised and each has its own number operator that commutes with the quantum Hamiltonian. This leads to the Fock representation of a quantum field in which eigenstates of the Hamiltonian are labelled by particle occupation numbers.

Because in reality we deal mainly with stable systems whose Hamiltonians have a lower bound, it is not so widely known that for unstable systems, there does not even exist a complete set of number operators that commute with the Hamiltonian. Under the action of the group of symplectic transformations, indefinite quadratic Hamiltonians have five types of irreducible canonical form beyond that of the harmonic oscillator. For the other classes, there is no vacuum state that can be defined as a zero-particle state. The Fock representation breaks down. The Hamiltonian has continuous spectrum, unbounded below, and a basis of solutions to the governing wave equation must have runaway non-oscillatory modes.

Minkowski space is only a local approximation. Boson fields must be coupled to the gravitational field of a universe that is expanding and accelerating. The fields may become unstable. When a scalar Klein-Gordon field is weakly coupled to the gravitational field of a spatially flat and exponentially stretching de Sitter space-time, the component building blocks of one dynamical degree of freedom are harmonic oscillators with time-dependent decreasing force constant that eventually will be negative. They may transition from being an attractive simple oscillator to a repulsive simple repeller. With transitions to instability occurring over time at successively shorter material wavelengths, the basic components of the field include a finite number of simple repellers that have pure imaginary angular frequencies. At a sequence of finite times, each solution mode ceases to oscillate. Working through the steps of a long-established and trusted approach [1], the quantum Hamiltonian consists of two parts; one is equivalent to the familiar infinite sum of commuting number operators, while the other is equivalent to a finite sum of pair creation operators and pair annihilation operators that cannot be transformed to number operators by a Bogoliubov transformation [2]. The energy eigenstates of the repellers are neither wave-like nor particle-like. They would not resonate with wave detectors or particle detectors. In that sense they are dark.

According to prevailing theoretical models and observations, currently the energy of the universe is predominantly dark energy, associated with exponential expansion. In its previous phase, cosmological mass-energy was predominantly due to matter, leading to an expansion that was of the $\frac{2}{3}$ degree in cosmic time. Until the early plasma phase, around 47,000 years after the big bang, most of the energy budget was filled by photon and neutrino radiation, leading to expansion in proportion to square root of time.

Due to a period of rapid inflation, ending around around 10^{-24}s after the big bang, much of the universe is beyond the horizon of light signals. Inflation explains why distant parts of space that have recently come into view, are still in thermal equilibrium with the microwave background. Because of the variable expansion rates, we are motivated to ask which expansion rate laws cause destabilisation of scalar and electromagnetic vector fields when they are coupled to Friedmann-LeMâitre-Robertson-Walker (FLRW) isotropic homogeneous expanding space-times. Following on from that, we plan to quantise the destabilised electromagnetic field.

For any FLRW expansion law, one may trace sufficiently far back in time to a field Hamiltonian that did commute with a complete set of number operators, and there was a well defined vacuum state of lowest energy. Having defined such a state, one may propagate it forwards in time. That is the general idea behind the Bunch-Davies vacuum [3]. For gentle expansion accelerations, the vacuum states at one time may have particles, as counted by number operators that commuted with the Hamiltonian at an earlier time. A basic example of relativity of particle number is given by the Unruh effect [4], in which the temperature of the particle spectrum is proportional to the constant acceleration of an observer. For some less gentle but reasonable expansion laws, the Bogoliubov transformation that relates different sets of creation and annihilation operators for single-particle basis wave functions appropriate at different times, will not be implementable by a unitary transformation. In those cases, the two Fock representations will not be physically equivalent and the energy of the vacuum may be infinite, as measured with the Hamiltonian of an earlier time. Forty years ago that phenomenon was investigated in Ref. [5]. Due to the degeneracy of energy spectral values of such symmetric fields, the number of unstable modes increases rapidly as the square of the spatial wave number of the destabilising mode [5]. It is then feasible that the energy density of the unstable modes may be approximately constant at large times [2].

Beyond asking whether vacuum states at two different times are related, the more fundamental problem that we are addressing here, is that the evolving Hamiltonian may eventually generate unstable field dynamics, in which case the Fock representation breaks down and the very existence of such a vacuum state is negated, at least for a subsystem of finite degrees of freedom.

2. Symplectic transformations and Bogoliubov transformations

The relationship between symplectic transformations and Bogoliubov transformations was reviewed and developed in Ref. [6]. Most dynamical systems in mathematical physics can be derived from a principle of least action,

$$\delta\int_a^b L(\mathbf{q};\dot{\mathbf{q}})dt = 0, \tag{1}$$

where L is the Lagrangian. If $\left\{q_j : j = 1\cdots N\right\}$ is a general coordinate system for configuration space, the conjugate momentum variables are $p_j = \partial L/\partial \dot{q}_j$. Note that the conjugate pairs $\left(q_j, p_j\right)$ need not have the standard measurement dimensions of displacement and lineal momentum. However prior to non-dimensionalisation, $q_j p_j$ should have the standard dimensions of action. Hamiltonian mechanics allows additional generalisation. Although $\left(q_j, p_k\right)$ must represent phase space, q_j need not represent configuration space. For example, stable systems should be transformable to a system that is invariant under Born reciprocity [7], $\left(Q_j, P_j\right) = \left(-p_j, q_j\right)$. Hamilton's equations of motion in phase space are

$$\dot{q}_j = \frac{\partial H}{\partial p_j} \tag{2}$$

$$\dot{p}_j = \frac{-\partial H}{\partial q_j};\ j = 1\cdots N, \tag{3}$$

where $H(\mathbf{q};\mathbf{p}) = \sum_1^N p_j\dot{q}_j - L$. As set out in Ref. [8], an equilibrium point of that system may be shifted to the origin and the Hamiltonian is then approximately quadratic $H = \frac{1}{2}z^t\hat{H}\mathbf{z} + \mathcal{O}\left(|\mathbf{z}|^2\right)$, where $\mathbf{z} = (\mathbf{q},\mathbf{p})^t$ and $\hat{H}$ is a real symmetric $2N \times 2N$ matrix. Hamilton's equations are preserved by canonical transformations $\left(q_j, p_j\right) \mapsto \left(Q_j(\mathbf{z}), P_j(\mathbf{z})\right)$ that preserve the Poisson bracket relations

$$\left\{Q_j, Q_k\right\} = \sum_{m=1}^{N} \frac{\partial Q_j}{\partial q_m}\frac{\partial Q_k}{\partial p_m} - \frac{\partial Q_j}{\partial p_m}\frac{\partial Q_k}{\partial q_m} = 0, \tag{4}$$

$$\left\{P_j, P_k\right\} = 0, \tag{5}$$

$$\left\{Q_j, P_k\right\} = \delta_{j,k}. \tag{6}$$

Linear canonical transformations $\mathbf{z} = C\mathbf{Z}$, are represented by real $2N \times 2N$ symplectic matrices that preserve a non-degenerate skew-symmetric form with Gram matrix G. $C^tGC = G$ with $G^t = -G$. Without loss of generality, $G = \begin{pmatrix} 0 & -I_N \\ I_N & 0 \end{pmatrix}$. Then the quadratic forms transform as $\hat{H}' = C^t\hat{H}C \Leftrightarrow G\hat{H}' = C^{-1}G\hat{H}C$. It is well known that a positive definite quadratic Hamiltonian can always be transformed to that of a system of independent harmonic oscillators with angular frequencies ω_j that are real eigenvalues of $iG\hat{H}$. Since $iG\hat{H}$ transforms by a similarity transformation, these angular frequencies are invariants. Among the canonical forms $G\hat{H}$ for conjugacy classes of generators of the symplectic group, there are five different types [9]. Their irreducible Jordan blocks may be arbitrarily large and they account for complex frequencies. A simple repeller with $N = 1$ and $H = \frac{1}{2}p^2 - \frac{1}{2}\alpha^2q^2$ has pure imaginary angular frequencies $\pm i\alpha$ so it cannot be transformed to a simple oscillator that has real angular frequencies.

2.1 Heisenberg algebra

In quantum mechanics, real functions on phase space are replaced by self-adjoint operators. Under the Dirac correspondence principle, the skew-symmetric commutator brackets between quantum observables q_j and p_k take the same values as for the Poisson brackets in (5) multiplying $i\hbar I$ where $2\pi\hbar$ is Planck's constant. These are the canonical commutation relations (CCR). Those commutator relations generate the Heisenberg Lie algebra and the corresponding unitary operators $e^{i(\mathbf{q},\mathbf{p})^t\mathbf{z}}; \mathbf{z} \in \mathbb{R}^{2N}$, are the building blocks of the Heisenberg-Weyl group.

Now it is convenient to rescale quantum phase space variables to non-dimensional form, choosing scales m_s, ℓ_s and t_s for mass, length and time so that $\hbar = 1$. Conforming with reference [2], we choose $\ell_s = R_0$ (universe radius at present time as reference), $t_s = R_0/c$ and $m_s = \hbar/cR_0$. This also scales c to 1, so that x_0 may be identified with t. When dealing with electromagnetic phenomena, there is one more independent quantity, such as charge, to be scaled. When dealing with free fields, it is more convenient to choose the electrical permittivity of the vacuum μ_0 as the permittivity scale. Since $c^2 = \mu_0 \varepsilon_0$ is scaled to 1, the magnetic permeability μ_0 automatically scales to 1 also.

2.2 Boson algebra

After non-dimensionalising, let $a_j = 2^{-1/2}\left(q_j + ip_j\right)$. Then a_j and their adjoints $a_j^\dagger$ satisfy the Boson CCR: $[a_j, a_k] = 0,\ \left[a_j^\dagger, a_k^\dagger\right] = 0,\ \left[a_j, a_k^\dagger\right] = \delta_{jk}$. Hence, quadratic Hamiltonians may be expressed as quadratic combinations of annihilation and creation operators. With unitary matrix $\mathcal{M} = 2^{-1/2}\begin{pmatrix} I_N & I_N \\ -iI_N & iI_N \end{pmatrix}$,

$$H = \frac{1}{2}\mathbf{z}^t \hat{H}\mathbf{z} = \frac{1}{2}\left(\mathbf{a}^\dagger, \mathbf{a}\right)\mathcal{D}\left(\mathbf{a}, \mathbf{a}^\dagger\right)^t, \tag{7}$$

where $\mathcal{D} = \mathcal{M}^\dagger \hat{H}\mathcal{M}$. Now $\hat{I}\mathcal{M}^\dagger = -i\mathcal{M}^\dagger G$, where $\hat{I} = diag[I_N, -I_N]$. It follows that $\hat{I}\mathcal{D} = \mathcal{M}^{-1}\left(-iG\hat{H}\right)\mathcal{M}$. By similarity, $\hat{I}\mathcal{D}$ must have the same eigenvalues as $-iG\hat{H}$, namely the characteristic frequencies $\pm\omega_j$. For example, the Hamiltonian of a harmonic oscillator may have $H = \frac{1}{2}P^2 + \frac{1}{2}\omega^2 Q^2$. After the canonical transformation $(Q, P) = \left(\omega^{-1/2}q, \omega^{1/2}p\right)$, $H = \frac{1}{2}\omega P^2 + \frac{1}{2}\omega Q^2 = \frac{1}{2}\omega a^\dagger a + \frac{1}{2}\omega a a^\dagger = \frac{1}{2}\omega + \omega a^\dagger a$. Whenever $\hat{H}$ (and therefore $\mathcal{D}$) is positive definite, the Hamiltonian may be expressed as a ground state energy plus a linear combination of commuting number operators $\sum_j \omega_j a_j^\dagger a_j$. This may require a canonical transformation or a Bogoliubov transformation to render the Hamiltonian in this standard form. A Bogoliubov transformation $\left(a_j, a_k^\dagger\right) \mapsto \left(b_j, b_k^\dagger\right)$ involves linear combinations of annihilation and creation operators that preserve the Boson CCR as well as Hermitian conjugacy between b_j and $b_j^\dagger$. In order to maintain the CCR, the linear transformation $\begin{pmatrix} \mathbf{a} \\ \mathbf{a}^\dagger \end{pmatrix} = \mathcal{T}\begin{pmatrix} \mathbf{b} \\ \mathbf{b}^\dagger \end{pmatrix}$ satisfies $\mathcal{T}^\dagger \hat{I}\mathcal{T} = \hat{I}$; i.e. it must be pseudo-unitary in $U(N, N)$. To satisfy the Hermitian conjugacy, it must have the block structure

$$\mathcal{T} = \begin{pmatrix} T_1 & T_2 \\ T_2^* & T_1^* \end{pmatrix}. \tag{8}$$

Because of the equivalence (7), the group of Bogoliubov transformations is isomorphic to the real symplectic group. It must preserve characteristic frequencies. For indefinite Hamiltonians, the characteristic frequencies need not be real. For example the indefinite self-adjoint Hamiltonian $H = \frac{1}{2}\left[i\alpha aa - i\alpha a^\dagger a^\dagger\right]$ has pure imaginary characteristic frequencies $\pm i\alpha$. It cannot be transformed to a linear combination of number operators which would have real frequencies. In that case, there can be no number operators that commute with the indefinite Hamiltonian [10]. There is no zero-particle vacuum state that would have any particular significance for that dynamical system. Energy eigenstates of the quantum Hamiltonian could not be labelled by particle numbers.

2.3 Indefinite metric on single mode space

If $iG\hat{H}$ has real eigenvalues $\pm\omega_k$, then there exists a canonical transformation that reduces the system to a collection of independent harmonic oscillators. The general solution of Hamilton's equations for a single harmonic oscillator with $H = \frac{1}{2}(p^2 + \omega^2 q^2)$ is

$$q = Ae^{-i\omega t} + A^* e^{i\omega t}; p = -i\omega Ae^{-i\omega t} + i\omega A^* e^{i\omega t}. \tag{9}$$

After quantisation, the amplitudes must be generalised to operators A and $A^\dagger$. It is well known that scaled amplitudes $A/\sqrt{2\omega}$ and $A^\dagger/\sqrt{2\omega}$ satisfy the Boson CCR for annihilation and creation operators. The classical solution space may be regarded as a single-particle space. This is behind the general idea of second quantisation; in that case, the wave function solution space for the classical or 'first quantised' field is the single particle space for the second-quantised field. In the Cook construction of a Fock representation, n-particle space is a n-fold tensor product of the classical solution space [11]. However for unstable systems with quadratic Hamiltonian, the CCR among operators q and p are incompatible with the Boson CCR among quantised amplitude operators A and $A^\dagger$. For example, consider a charged repeller with $H = p^\dagger p - q^\dagger q = \frac{1}{2}\left[\left(p_1^2 + p_2^2\right) - \left(q_1^2 + q_2^2\right)\right]$, where $q = 2^{-1/2}(q_1 - iq_2)$ and $p = 2^{-1/2}(p_1 + ip_2)$. The general solution is of the form

$$q = \sqrt{2}Ae^{-t} + \sqrt{2}Be^{t};\ \ p = -\sqrt{2}A^\dagger e^{-t} + \sqrt{2}B^\dagger e^{t}. \tag{10}$$

If we demand that the mode amplitude operators satisfy the Boson CCR, $\left[A, A^\dagger\right] = I$, $\left[B, B^\dagger\right] = 1$ and other commutators vanish, then it follows that

$$\left[q_1, q_2\right] = -2i\cosh(2t). \tag{11}$$

Similar commutators appear among field operators $\Phi(x^\mu)$ and $\Phi(y^\mu)$ at space-like separation $x^\mu - y^\mu$ when the field obeys a Klein-Gordon equation with $m^2 < 0$. Such a field may be decomposed as a system of independent oscillators and independent repellers. In Ref. [12], the imaginary-mass Klein-Gordon equation was quantised using an approach that is opposite to the above. If one assumes standard CCR among the operators q and p, then it follows that

$$\left[A, B^\dagger\right] = \frac{1}{2}i. \tag{12}$$

whereas other commutators vanish. If one assumes the existence of a normalised cyclic vector ψ_0 in a Hilbert space, satisfying $A\psi_0 = 0$ and $B\psi_0 = 0$, then there follows the inner product $\langle A^\dagger\psi_0|B^\dagger\psi_0\rangle = \frac{1}{2}i$. In the basis $\{A^\dagger\psi_0, B^\dagger\psi_0\}$, single-mode complex Hilbert space has a sesquilinear inner product with Gram matrix $\begin{pmatrix} 0 & i/2 \\ -i/2^* & 0 \end{pmatrix}$, which is indefinite with eigenvalues $\pm\frac{1}{2}$. An indefinite inner product arose in the Gupta-Bleuler method to quantise the electromagnetic field [13]. However in flat Minkowsi space, the dynamics maintain a separation between physical states of positive norm and unphysical states of negative and zero norm, so that the latter may be factored out. This occurs when the basic components are of the form

$H = \frac{1}{2}(p_1^2 + \omega_1^2 q_1^2) - \frac{1}{2}(p_2^2 + \omega_2^2 q_2^2)$. Then the component with positive definite energy commutes with the total Hamiltonian. In all other cases of indefinite quadratic Hamiltonian, this is no longer true [14].

3. Electromagnetic field

3.1 From finite to infinite degrees of freedom

Field variables $\Phi(x,t)$ take time-dependent values at each point in space. Therefore they have infinite degrees of freedom, often labelled by space location x. Most of the fields of mathematical physics obey an action principle

$$\delta \int \int_{\Omega} \mathcal{L}(\Phi(x,t); \Phi,_{\mu}) dV\, dt = 0, \tag{13}$$

where the Lagrangian density $\mathcal{L}$ is meant to depend on Φ as well as its time derivative $\Phi,_0 = \partial\Phi/\partial x^0$ and space derivatives $\Phi,_j = \partial\Phi/\partial x^j; j = 1,2,3$. Here Ω is the spatial domain. The canonical conjugate field is $\Pi(x,t) = \partial\mathcal{L}/\partial\Phi,_0$. Provided $\mathcal{L}$ depends on $\Phi,_0$, canonical quantisation may proceed in the usual way. Equal-time canonical commutation relations now extend the Kronecker delta to Dirac delta function: $[\Phi(x,t), \Pi(y,t)] = i\delta(\mathbf{x} - \mathbf{y})$. The dynamical field equations are the Euler-Lagrange equations that follow from the action principle:

$$\frac{\partial\mathcal{L}}{\partial\Phi} - \partial_\lambda \frac{\partial\mathcal{L}}{\Phi,_\lambda} = 0. \tag{14}$$

In relativistic formulations we will assume the Einstein summation convention that Greek indices repeated at lower (covariant) and upper (contravariant) levels, will be summed from 0 to 3, whereas repeated Roman indices will be summed from 1 to 3. It will be assumed that space-time is a pseudo-Riemannian manifold with covariant symmetric metric tensor $g_{\mu\nu}$ with metric defined locally by $(\delta s)^2 = g_{\mu\nu} dx^\mu dx^\nu$. Contravariant components $g^{\mu\nu}$ of the metric tensor are defined to be the matrix entries of the inverse of $g_{\mu\nu}$; $g^{\mu\nu} g_{\nu\lambda} = \delta^\mu_\lambda$.

Having obtained a basis $\{f_k\}$ of solutions $f_k(x,t)$ of the field equations, we may re-express the quantised Hamiltonian in terms of operators a_k that satisfy Boson CCR. For fields in Minkowski space the labels k may be wave vectors in the Fourier basis, such that all Boson CCR including $\left[a_{\mathbf{k}}, a^\dagger_{\mathbf{p}}\right] = \delta(\mathbf{k} - \mathbf{p})$, are satisfied. On bounded domains, the basis is countable and the commutators are zero or Kronecker delta.

Just as in quantum mechanics with finite degrees of freedom, field theories may be described in terms of self-adjoint quantum fields [15], or alternatively in terms of creation and annihilation operators [16]. A Bogoliubov transformation is implementable as a unitary transformation $b_k = Ua_k$ in a Hilbert space representation of the quantum field, if and only if the off-diagonal creation-to-annihilation block T_2 in (8), is of the Hilbert-Schmidt class, i.e. trace of $T_1^\dagger T_1$ is finite. Many classical field models are Hamiltonian, leaving invariant some symplectic form between pairs of solutions [17]. An infinite dimensional symplectic transformation C is unitarily implementable if and only if $C^t C - I$ is of the Hilbert-Schmidt class [15]. These two

Hilbert-Schmidt conditions are equivalent [6]. The one-parameter symplectic flow given by unstable field dynamics, need not be unitarily implementable. Then in the Fock representation, the vacuum state for a set of evolving annihilation operators may diverge in the norm. The rate of change of energy of an evolving vacuum state may diverge [18]. In the context of scalar Boson fields coupled to a FLRW space-time, this problem can be avoided only through conformal coupling [5].

3.2 Electromagnetic 4-potential

The free-field Maxwell field equations are encapsulated as the zero divergence of the Faraday electromagnetic tensor, along with the Bianchi identity:

$$\nabla_\nu F^{\mu\nu} = 0, \tag{15}$$

$$\partial_\mu F_{\alpha\beta} + \partial_\alpha F_{\beta\mu} + \partial_\beta F_{\mu\alpha} = 0. \tag{16}$$

Here, ∇_μ is the covariant gradient.

3.2.1 Contravariant components and physical components

Consider a vector $\mathbf{A} = A^\mu \mathbf{e}_\mu$ in terms of a non-normalised basis in an indefinite inner product space. Let $g_{\mu\nu} = \mathbf{e}_\mu \cdot \mathbf{e}_\nu$. The 'physical' components of a vector are traditionally taken with respect to a normalised basis $\mathbf{e}'_\mu = \mathbf{e}_\mu / \sqrt{|g_{\mu\mu}|}$. Now $\mathbf{A} = A'^\mu \mathbf{e}'_\mu$ with $A'^\mu = \sqrt{|g_{\mu\mu}|} A^\mu$. Then $\mathbf{A} \cdot \mathbf{A} = A^\mu A^\nu g_{\mu\nu} = A'^\mu \eta_{\mu\nu} A'^\nu$, where $\eta_{\mu\nu}$ is the normalised metric tensor $g_{\mu\nu}/\sqrt{|g_{\mu\mu} g_{\nu\nu}|}$. In the current application, this is just the Minkowski metric.

Similarly, components of the tensor $F^{\mu\nu}$ may be normalised to physical components $F'^{\mu\nu} = F^{\mu\nu}\sqrt{|g_{\mu\mu} g_{\nu\nu}|}$.

The matrix entries of skew-symmetric $F'^{\mu\nu}$ are components of electric and magnetic fields: $F'^{\mu\nu} = \begin{pmatrix} 0 & -E^1 & -E^2 & -E^3 \\ E^1 & 0 & -B^3 & B^2 \\ E^2 & B^3 & 0 & -B^1 \\ E^3 & -B^2 & B^1 & 0 \end{pmatrix}$.

The field equations are also written in terms of the electromagnetic 4-vector potential so that $F_{\mu\nu} = \partial_\mu A_\nu - \partial_\nu A_\mu$. Because of the skew symmetry of that expression, if the space is torsionless, then the covariant derivatives are not required in this equation as the connection terms cancel.

It is easily verified that

$$-\frac{1}{4} F^{\mu\nu} F_{\mu\nu} = -\frac{1}{4} F'^{\mu\nu} \eta_{\mu\alpha} \eta_{\nu\beta} F'^{\alpha\beta} = \frac{1}{2} \sum_j E'^j E'^j - B'^j B'^J. \tag{17}$$

This is the usual Lagrangian density of classical electrodynamics. There may be different choices of gauge by which different potential fields $A_\nu(x,t)$ lead to the same physical fields E_j and B_j. The simplest type of wave equation for A_μ is sometimes satisfied after choosing the Lorenz gauge (a particular form of Lorentz gauge),

$$\nabla^{\mu} A_{\mu} = 0. \tag{18}$$

Then in flat Minkowski space, the free field equations $\partial_{\nu} F^{\mu\nu} = 0$ reduce to the linear wave equation $\partial_{\mu}\partial^{\mu} A_{\nu} = \partial_0^2 A_{\nu} - \nabla^2 A_{\nu} = 0$. In extensions of the field equations to a curved pseudo-Riemannian manifold, it is natural to replace the 4D Laplacian by the Laplace-Beltrami operator. However it is not known exactly how the electromagnetic field couples to the curvature. When the covariant derivatives are allowed to operate on general tensor fields, ∇_{μ} and ∇_{ν} do not necessarily commute. In particular, $[\nabla_{\mu}, \nabla_{\nu}] A^{\mu} = R_{\nu\alpha} A^{\alpha}$, where $R_{\mu\nu}$ is the symmetric Ricci tensor. The consequent ambiguity in extending the flat-space Laplacian is covered by allowing such coupling terms to be forcing terms in the field eqs. [19]. If, as in the theory of scalar fields, one also considers coupling to the Ricci scalar R, then the field equations are of the form

$$\frac{1}{\sqrt{|g|}} \partial_{\mu} \left(\sqrt{(|g|)} g^{\mu\lambda} \partial_{\lambda} A^{\nu} \right) = \zeta g^{\nu\lambda} R_{\alpha\lambda} A^{\alpha} + \xi R A^{\nu}, \tag{19}$$

where $g = det\{g_{\mu\nu}\}$. This breaks gauge invariance when $a(t)$ is not static. However we use it as an indicative model to examine qualitative consequences of electromagnetic-gravitational interactions. Some classical fields of this type were developed in Ref. [20]. We show below, that even with minimal coupling, the model wave equations are related to the flat-space Proca equation with time dependent mass term. For the standard Proca system with mass, the Lorenz gauge condition is not only a convenient choice for simplifying the equations but it is a necessary consequence of Proca's covariant formulation [21]. In dimensionless terms, the Lagrangian density is of the form

$$\mathcal{L} = \frac{-1}{4} F_{\alpha\beta} F^{\alpha\beta} \sqrt{(|g|)} + \frac{1}{2} \zeta R_{\alpha\beta} A^{\alpha} A^{\beta} + \frac{1}{2} \xi R A^{\alpha} A_{\alpha}. \tag{20}$$

In the case of minimal coupling ($\zeta = 0$ and $\xi = 0$), the theory is gauge invariant since the Lagrangian density is expressible in terms of $F_{\mu\nu}$. Otherwise, the theory is not gauge-invariant over cosmological scales and the 4-potential will be viewed as a measurable field. However numerical values of the coupling constants are not known. Another popular model is conformal coupling $\xi = \frac{1}{6}$ and $\zeta = 0$. Independently of the form of $a(t)$, that allows the field equations to be transformed to the standard vector-valued wave equation by a change of variables. However we show here that a special case of the expansion factor, namely the exponential case that our universe is approaching, allows other conformal combinations

$$\xi = \frac{1}{6} - \frac{\zeta}{4}. \tag{21}$$

This includes the simple case $\xi = 0$ and $\zeta = \frac{2}{3}$.

3.3 Restriction to spatially flat FLRW universes

Apart from relative deviations of less than 10^{-4} observed in cosmic microwave background temperature, at cosmological scales the universe is approximately homogeneous and isotropic, with flat spatial cross sections. Since the 1970s there has been considerable development of the quantum theory of scalar fields coupled to FLRW expanding space-time [22, 23].

When analysing dynamical fields, we choose to use the convention that time-like separations are positive. The homogeneous isotropic metric is then of the form

$$(ds)^2 = (dx^0)^2 - a(x^0)^2\left[(dx^1)^2 + d(x^2)^2 + (dx^3)^2\right] \tag{22}$$

Here, $x^j;\ j = 1\cdots3$, are material coordinates that remain constant on a fixed material point that is moving with the average cosmological expansion. t is cosmic time as measured by an observer attached to such a material point. $a(t)$ is the expansion factor for stretching space. In the FLRW form of (19), we need exact expressions for the Ricci scalar and the Ricci tensor. These are

$$R_{\mu\nu} = \mathrm{diag}\left[-3\frac{a''(t)}{a}, 2(a')^2 + aa'', 2(a')^2 + aa'', 2(a')^2 + aa''\right)\Bigg], \tag{23}$$

$$R = R^\mu_\mu = -6\left[\left(\frac{a'}{a}\right)^2 + \frac{a''}{a}\right]. \tag{24}$$

For general $a(t)$, the field Eq (19) takes the form

$$\partial_t^2 A^0 + 3\frac{a'(t)}{a}\partial_t A^0 - a^{-2}\sum_{k=1}^{3}\partial_k\partial_k A^0 + (6\xi + 3\zeta)\frac{a''}{a}A^0 + 6\xi\left(\frac{a'}{a}\right)^2 A^0 = 0, \tag{25}$$

$$\partial_t^2 A^j + 3\frac{a'(t)}{a}\partial_t A^j - a^{-2}\sum_{k=1}^{3}\partial_k\partial_k A^j + (6\xi + \zeta)\frac{a''}{a}A^j + [6\xi + 2\zeta]\left(\frac{a'}{a}\right)^2 A^j = 0, \tag{26}$$

for $j = 1,2,3$. For the case of exponential expansion with present time at $t = 0$, $a(t) = e^{\sqrt{\frac{\Lambda}{3}}t}$, $R_{00} = -\Lambda$, $R_{11} = R_{22} = R_{33} = \Lambda a(t)^2$ and $R = -4\Lambda$.

For the case of power-law expansion with present time at $t = 1$, $a(t) = t^\alpha$, $R_{00} = -3\alpha(\alpha - 1)t^{-2}$, $R_{11} = R_{22} = R_{33} = \alpha(3\alpha - 1)t^{2(\alpha-1)}$ and $R = 6\alpha(1 - 2\alpha)t^{-2}$.

3.4 Conformal time and conformal coupling

Aiming to quantise the field equations, the coordinate system should be chosen so that the field equations are as close as possible to Lorentz-invariant wave equations for which the canonical quantisation method is well established. Therefore it is standard practice to adopt the conformal time coordinate η with t_0 chosen so that the current time is either $t = 1$ or $t = 0$:

$$\eta = \int_{t_0}^{t}\frac{1}{a(t_1)}dt_1. \tag{27}$$

Then the metric conforms to that of Minkowski but with non-zero curvature invariants:

$$ds^2 = a(t(\eta))^2\left(d\eta^2 - (dx^1)^2 - (dx^2)^2 - (dx^3)^2\right). \tag{28}$$

The leading cosmological paradigm [24] is that of Lambda-Cold Dark Matter (ΛCDM). From soon after inflation, up to age 47,000 years, the energy was

a(t)	η(t)	a(t(η))	Range η
$(t+1)^{\alpha}; 0<\alpha<1$	$\frac{(t+1)^{1-\alpha}-1}{1-\alpha}$	$[1+(1-\alpha)\eta]^{\alpha/(1-\alpha)}$	$\left(\frac{-1}{1-\alpha},\infty\right)$
$t+1$	$\log(t+1)$	e^{η}	$(-\infty,\infty)$
$(t+1)^{\alpha}; \alpha>1$	$\frac{1-(t+1)^{-(\alpha-1)}}{\alpha-1}$	$[1-(\alpha-1)\eta]^{-\alpha/(\alpha-1)}$	$\left(-\infty,\frac{1}{\alpha-1}\right)$
$e^{t\sqrt{\Lambda/3}}$	$\sqrt{\frac{3}{\Lambda}}\left[1-e^{-t\sqrt{\Lambda/3}}\right]$	$\frac{\sqrt{3/\Lambda}}{\sqrt{\frac{3}{\Lambda}}-\eta}$	$\left(-\infty,\sqrt{3/\Lambda}\right)$

Table 1.
Expansion factor in terms of conformal time η.

predominantly in the form of photon and neutrino radiation. During that period the dominant term in $a(t)$ was proportional to $t^{1/2}$. For most of the duration of the plasma universe which lasted up to the time of atomic recombination at age 360,000 years, the mass-energy was already dominated by matter. The time of last scattering of photons occurred around 380,000 years after the big bang. During the matter-dominated epoch, up to age 9.8×10^{9} years, the dominant term in $a(t)$ was proportional to $t^{2/3}$. Since then, the expansion has been predominantly exponential $a(t)\sim e^{t\sqrt{\Lambda/3}}$ where Λ is the cosmological constant, proportional to dark energy density. Power-law and exponential expansion factors, converted to functions of η, are shown in **Table 1**.

After changing time coordinate t to η, the field equations are, with over-dot denoting η-derivative,

$$\partial_\eta^2 A^0-\sum_{k=1}^{3}\partial_k\partial_k A^0+2\frac{\dot a(\eta)}{a}\partial_\eta A_0+6\xi\frac{\ddot a}{a}A^0+3\zeta\left[\frac{\ddot a}{a}-\left(\frac{\dot a(\eta)}{a}\right)^2\right]A^0=0, \tag{29}$$

$$\partial_\eta^2 A^j-\sum_{k=1}^{3}\partial_k\partial_k A^j+2\frac{\dot a(\eta)}{a}\partial_\eta A_j+6\xi\frac{\ddot a}{a}A^j+\zeta\left[\frac{\ddot a}{a}+\left(\frac{\dot a(\eta)}{a}\right)^2\right]A^j=0. \tag{30}$$

3.4.1 Amplifying the signal

In the partial differential equation for any of the components A^μ, the first two terms are those of the standard d'Alembert wave equation. However the next term is a first-order damping term, again in common with each component equation. That term may be transformed away by changing the dependent variable to the amplified 4-potential $\overline{A}^\mu=a(t(\eta))A^\mu$. Note that $\overline{A}^\mu$ is still a *bona fide* observable vector quantity. Its evolution equation still is derivable from a Lagrangian density and its quantum counterpart and the conjugate variables will still satisfy the CCR. The governing Eq. (30) is now expressed as

$$\partial_\eta^2\overline{A}^0-\sum_{k=1}^{3}\partial_k\partial_k\overline{A}^0+\left([-1+6\xi+3\zeta]\frac{\ddot a}{a}-3\zeta\left[\frac{\dot a}{a}\right]^2\right)\overline{A}^0=0, \tag{31}$$

$$\partial_\eta^2\overline{A}^j-\sum_{k=1}^{3}\partial_k\partial_k\overline{A}^j+\left([-1+6\xi+\zeta]\frac{\ddot a}{a}+\zeta\left[\frac{\dot a}{a}\right]^2\right)\overline{A}^j=0. \tag{32}$$

3.4.2 Dynamical permutability, conformality and instability

We say that the dynamical equations for 4-vector components $\overline{A}^0$ and $\overline{A}^j$ are *permutable* if each component satisfies the same equation. Otherwise the components would be distinguishable by dynamical behaviour alone. Assume $a(\eta)$ is increasing and $a(0) = 1$. From (32), dynamical permutability is equivalent to

$$\zeta\left[\left(\frac{\ddot{a}}{a}\right) - 2\left(\frac{\dot{a}}{a}\right)^2\right] = 0 \tag{33}$$

$$\Leftrightarrow \zeta = 0 \ \text{ or } \left(\frac{\ddot{a}}{a}\right) = 2\left(\frac{\dot{a}}{a}\right)^2 \tag{34}$$

$$\Leftrightarrow \zeta = 0 \ \text{ or } a = \frac{\eta_\infty}{\eta_\infty - \eta}, \tag{35}$$

with η_∞ constant. Referring to **Table 1**, we see that this dependence of a on η is equivalent to the expansion factor a being exponential in cosmic time, with Hubble time $t_H = a/\dot{a} = \sqrt{3/\Lambda} = \eta_\infty$. Therefore we have the result:

A dynamically permutable 4-vector potential can be coupled to the Ricci tensor ($\zeta \neq 0$) if and only if $a(t)$ is exponential in time.

Since the expansion factor approaches the exponential asymptotically, it may be considered that ζ is non-zero, and that the earlier radiation-dominated epoch and mass-dominated epoch allowed for relatively short-lived dynamically inconsistent stages in the long-term development of the universe.

In dynamically permutable models, the vector field equation reduces to a Proca equation wherein each component satisfies the same Klein-Gordon equation with time-dependent squared mass:

$$\partial_\eta^2 \overline{A}^\mu - \sum_{k=1}^{3} \partial_k \partial_k \overline{A}^\mu + M^2(\eta)\overline{A}^\mu = 0. \tag{36}$$

Effectively, the electromagnetic field equations on a curved FLRW space-time have been transformed to a Proca vector equation on flat-space-time but with time-dependent squared mass. The case of conformal coupling is that which reduces the PDEs to the standard linear wave equation with zero mass. For non-exponential expansion, this requires $\zeta = 0$ and $\xi = \frac{1}{6}$. This conformal value of ξ is well known from scalar field theory. For exponential expansion, the conformal condition is $\xi = \frac{1}{6} - \frac{1}{4}\zeta$, which includes the case $\xi = 0$ and $\zeta = \frac{2}{3}$.

The following cases obey vector Proca-Klein-Gordon equations with positive time-dependent squared mass: (i) $a(t) = (1+t)^\alpha; \alpha > \frac{1}{2}, \alpha \neq 1, \zeta = 0$ and $\xi > \frac{1}{6}$, (ii) $a(t) = (1+t)^\alpha; 0 < \alpha < \frac{1}{2}, \zeta = 0$ and $\xi < \frac{1}{6}$, and (iii) $a(t) =$ exponential and $\xi > \frac{1}{6} - \frac{1}{4}\zeta$.

Cabral and Lobo derived a wave equation with $\zeta = 1$ and $\xi = 0$, which is dynamically stable with $a(t)$ exponential.

The less studied and less understood models are those such as the minimal coupling model ($\zeta = 0, \xi = 0$) in which $\xi < \frac{1}{6} - \frac{1}{4}\zeta$, with power law $\alpha > \frac{1}{2}$ or exponential expansion so that $M^2(\eta) < 0$. In those cases, the Hamiltonian is unbounded below, the classical solutions have some non-oscillatory modes and some energy eigenstates of the quantum Hamiltonian are not eigenstates of a number operator that commutes with the Hamiltonian. That will be the case that we consider further here. The model with exponential $a(t)$ has

$$M^2 = [12\xi + 3\zeta - 2](\eta_\infty - \eta)^{-2}, \tag{37}$$

with $\eta_\infty = \sqrt{3/\Lambda}$. Note that with minimal coupling, the power law with very large α also is consistent with $M^2 = -2(\eta_\infty - \eta)^{-2}$ which results from the exponential $a(t)$. From the perspective of destabilisation, critical cases occur at $\alpha = 0$ (static), $\alpha = \frac{1}{2}$ (radiation-dominated), $\alpha = 1$ (Milne, equivalent to static) and $\alpha \to \infty$ (Λ-dominated). The case $\alpha = \frac{2}{3}$ (matter-dominated) has an additional special property that $\ddot{a}$ is constant.

4. Field quantization

From our first introduction to dynamics, we learn that models may become mathematically more tractable when we introduce additional variables that have some redundancy. The mechanics of a particle in a conservative force field is conceptually simpler after we introduce a potential function that is not uniquely defined but allows us to construct a first integral, the energy. This idea extends to the electromagnetic 4-potential. It allows us to collapse Maxwell's equations to a simpler mathematical form in terms of a set of potential variables with some redundancy. In Minkowski space, the dynamical equation for the 4-potential has gauge invariance $A_\mu \to A_\mu + \psi,_\mu$. A specific choice of gauge can be designed to facilitate a specific purpose, such as reduction to a standard wave equation which follows directly in the Lorenz gauge (18). At the same time, a specific choice of gauge removes some of the redundancy. Even after choosing a gauge, it first appears that there are three degrees of freedom at each point in space-time, when we know that there are two independent physical polarisations of electromagnetic waves. The extra redundancy is associated with the fact that the Lagrangian density (16), whether or not the Proca mass term is included, does not depend explicitly on $A^0,_0$. A conjugate momentum variable Π_0 cannot be defined. Consequently, canonical quantisation cannot proceed in the normal way. These comments are made in view of the historical context of extending the Proca system to a gauge theory, which eventually led to first-class constraints, a unitary representation of the commutation algebra on a positive normed space, and to renormalizability.

Before we consider quantising the non-autonomous Proca system (36), as a starting point, we can refer to the long and complicated history of the quantised Proca field with constant positive mass.

4.1 Standard autonomous Proca field

The standard Proca equation describes a vector Boson field with constant mass, in which case the components A^μ are not potentials but physical fields. The Proca equation originally appeared in the form

$$\partial_\nu F^{\nu\mu} + m^2 A^\mu = 0. \tag{38}$$

By taking ∂_μ of each side, by the skew-symmetry of $F^{\nu\mu}$, the first term is annulled and we are left with the Lorenz condition (e.g. [25]). The Proca equation is not gauge-invariant in the sense of electromagnetism. The Lorenz condition still must be imposed as a constraint. This is a troublesome second class constraint [26], formally $\mathcal{W}[\mathbf{A}] = 0$, although $\mathcal{W}$ does not commute through Poisson brackets with a naively constructed Hamiltonian. After quantisation it is not clear how a complete set of physical eigenstates of a Hamiltonian could belong to the zero-eigenspace of $\mathcal{W}$ and of its Hermitian conjugate as an operator in the Hilbert space representation. Since the earliest formulations of quantum vector field theory, a large body of work in mathematical physics has been devoted to various pathways to quantise systems with second class constraints. In the language of algebraic quantisation, this may be cast as the existence of regular states on the CCR algebra of observables that incorporates supplementary conditions [27]. In limited space, it would be difficult to do justice to a literature survey of the various concrete approaches that have been developed. They are described in the comprehensive review in Ref. [21]. The two most common approaches can be classified as (i) extending classical Poisson brackets to Dirac's constraint brackets before quantisation [26], and (ii) introducing one or more new fields that allow a new choice of gauge to change the second class constraints to first class constraints. The additional Stückelberg field ρ was introduced in 1938 [28] in the Lagrangian:

$$\mathcal{L} = -\frac{1}{2}\partial_\mu A^\nu \partial^\mu A_\nu + \frac{1}{2}m^2 A^\mu A_\mu + \frac{1}{2}\partial_\mu \rho \partial^\mu \rho - \frac{1}{2}m^2 \rho^2. \tag{39}$$

This allowed a new type of gauge invariance [1]:

$$A_\mu \mapsto A_\mu + \partial_\mu \chi, \tag{40}$$

$$\rho \to \rho + m\chi, \tag{41}$$

$$\partial_\mu \partial^\mu \rho + m^2 \rho = 0, \tag{42}$$

$$\partial_\mu \partial^\mu \xi + m^2 \chi = 0. \tag{43}$$

Instead of the Lorenz condition, one may impose a first-class constraint that involves all of the field variables. $\mathcal{L}$ now depends on $\partial_0 A^0$ so a conjugate momentum variable has the familiar definition $\Pi^\mu = \partial_0 \partial \mathcal{L}/\partial A^\mu_{,0}$. A Hamiltonian can be constructed. Integrating over a spatial section Ω,

$$H = \frac{1}{2}\int_\Omega - \left[\left(\Pi^0\right)^2 + m^2 \left(A^0\right)^2 + \nabla A^0 \cdot \nabla A^0 \right] + \sum_{j=1}^{3} \left(\Pi^j\right)^2 + m^2 \left(A^j\right)^2 + \nabla A^j \cdot \nabla A^j$$
$$+ \left[\left(\Pi^5\right)^2 + m^2 \rho^2 + \nabla \rho \cdot \nabla \rho \right] dV, \tag{44}$$

where Π^5 is the momentum variable conjugate to ρ. The first part with index 0, is negative and unbounded. However, as in the Gupta-Bleuler quantisation of the electromagnetic field, this was associated with unphysical states of negative norm, dynamically unconnected to physical states of positive norm. In general, Lorentz covariant locally gauge invariant fields must be represented on an indefinite inner product space [29]. Note that the matrix elements of the constraint operators showed zero mappings from the physical subspace to the unphysical subspace. Importantly, in the Stückelberg-Feynman gauge this then led to the same equal-time commutation relations among $A^\mu(x)$ and $\Pi^\mu(y)$ as in standard elecftrodynamics. With a non-zero time difference between operators, the commutators involved one solution of the massive Klein-Gordon equation.

The historical extensions of Stückelberg's formulation and its more recent ramifications, are described in Ref. [25]. Neglecting an additional inconsequential term that is a total derivative, the Lagrangian may be re-expressed

$$\mathcal{L} = -\frac{1}{4}F_{\mu\nu}F^{\mu\nu} + \frac{1}{2}m^2\left(A_\mu - \frac{1}{m}\partial_\mu\rho\right)\left(A^\mu - \frac{1}{m}\partial^\mu\rho\right) - \frac{1}{2}\left(\partial_\mu A^\mu + m\rho\right)^2. \tag{45}$$

The mass parameter is seen to be a coupling constant between the physical field and the gauge field. In this regard, Stückelberg's construction is seen to be a forerunner to the Higgs mechanism. In the Feynman-Stückelberg gauge, the last term can be equated to zero at the end of calculations. This allowed Feynman to quantise the field theory via path summations. Renormalisability of more general fields eventually followed through appropriate gauge fixing after the introduction of one extra real parameter along with a Faddeev-Popov Fermionic ghost field [30].

4.2 Proca equation with time-dependent mass

We are particularly interested in the case that the mass parameter in (36) is negative and decreasing in time. This includes the case of minimal coupling: (36) may still be derived from a real-valued Lagrangian density that is (39) with A^μ replaced by $\overline{A}^\mu$, m^2 replaced by $M^2(\eta)$ and ρ set to zero. In (45), the second term is no longer invariant under the Pauli gauge symmetry since a change of gauge introduces unbalanced terms that are quadratic in $\dot{M}$. At this stage, a gauge theory with time-dependent mass has not been developed. However the constraints and gauge conditions are treated, it remains true that each component A^μ satisfies a Klein-Gordon equation with negative mass that is decreasing in time. In the case of a spatially flat universe that is expanding exponentially or by power-law in cosmic time, M^2 is proportional to $-(\eta_\infty - \eta)^{-2}$. In those cases, with homogeneous Dirichlet or Neumann boundary conditions, the complete set of classical solutions of the non-autonomous Klein-Gordon equation can be constructed by separation of variables [2]. In terms of spherical harmonic functions of spherical polar coordinates and Bessel functions of r,

$$\overline{A}^\mu = \sum_{n=1}^{\infty}\sum_{l=0}^{2n}\sum_{m=-l}^{l}\frac{J_{l+\frac{1}{2}}(k_{nl}\,r)}{\sqrt{k_{nl}\,r}}Y_l^m(\theta,\varphi)\left\{a^\mu{}_{nlm}f_{n\ell}(\eta) + (-1)^m a^{\mu\dagger}{}_{nl-m}f^*_{n\ell}(\eta)\right\}. \tag{46}$$

The time dependent factors satisfy the equation of an oscillator with time dependent angular frequency $\omega_k = \sqrt{k^2 - 12\gamma(\eta_\infty - \eta)^{-2}}$:

$$\ddot{f}_k + \left[k^2 - 12\gamma(\eta_\infty - \eta)^{-2}\right]f_k = 0;\ \ \gamma = -\xi + \frac{1}{6}(1-\zeta). \tag{47}$$

The solutions are linear combinations of Bessel functions

$$(\eta_\infty - \eta)^{1/2} J_{\sqrt{12\gamma+1/4}}(k[\eta_\infty - \eta]) \text{ and } (\eta_\infty - \eta)^{1/2} Y_{\sqrt{12\gamma+1/4}}(k[\eta_\infty - \eta]). \tag{48}$$

Bessel functions of the second kind are ruled out because of their singularity at $\eta = \eta_\infty$. Then the solution reaches its last extremum at time $\eta = \eta_\infty - \lambda'_{\nu,0}/k_{n\ell}$, where $\lambda'_{\nu,0}$ is the smallest positive local extremum point of the Bessel function of order $\nu = \sqrt{12\gamma + 1/4}$. The subsequent period of time up to $\eta = \eta_\infty$ is of infinite duration in terms of cosmic time.

The un-amplified signal is $A^\mu = \overline{A}^\mu / a(\eta)$. When the order of the Bessel function is real, A^μ has temporal amplitude of the order $(\eta_\infty - \eta)^\kappa$ with $\kappa = \frac{3}{2} - \sqrt{\frac{9}{4} - 2\zeta + 12\xi}$. This amplitude is constant in the case of minimal coupling, or more generally when $\zeta + 6\xi = 0$.

Here, we have assumed a domain that is not elliptoid but spherical $r \leq 1$. This applies in a reference frame in which the cosmic background radiation is isotropic. Anisotropic redshifts show that our galaxy is moving at approximate speed 670 km s^{-1} relative to that frame [31]. Unlike in infinite Minkowski space, the wave numbers k take a countable spectrum of values $k_{n\ell}$. In unstable cases such as minimal coupling, each angular frequency decreases in time until it becomes zero and then purely imaginary. These attraction-repulsion transitions take place at later times for successively higher wave numbers. Each mode oscillates only a finite number of times in the future beyond any finite starting time.

When the solutions are known, calculation of the magnitude squared gradients in the Hamiltonian can be considerably simplified. Green's identity establishes

$$\int_\Omega \phi \nabla^2 \psi + \nabla\phi \cdot \nabla\psi dV = \int_{\partial\Omega} \phi \hat{n} \cdot \nabla\psi dS \tag{49}$$

where $\partial\Omega$ is the spatial boundary. Then with $\psi = \phi$, if we consider homogeneous boundary conditions of either Dirichlet or Neumann type, the boundary integral vanishes. Consequently

$$\frac{1}{2}\int_\Omega \left(A^\mu_{,\eta}\right)^2 + \nabla A^\mu \cdot \nabla A^\mu + M^2 (A^\mu)^2 dV = \frac{1}{2}\int_\Omega \left(A^\mu_{,\eta}\right)^2 - A^\mu A^\mu_{,\eta\eta} dV \tag{50}$$

In terms of material coordinates, we prefer the Neumann condition because in the usual physical coordinates, the stress-energy tensor then indicates zero transport of 4-momentum across the moving fluid material boundary. That boundary condition leads to the eigenvalues $k_{n\ell}$ being zeros of derivative spherical Bessel function j'_ℓ for any $\ell = 0,1,2,\cdots$. With Dirichlet boundary conditions, they would be zeros of j_ℓ. The spacing and interlacing of those eigenvalues are similar in the two cases. At large n, $k_{n\ell} \sim N\pi$.

A direct calculation [2] shows that the Hamiltonian at each frozen time is the sum of a repulsive component with finite but increasing degrees of freedom, plus the usual oscillatory component with infinite degrees of freedom. The number of repulsive modes is the number of independent modes that have pure imaginary angular frequencies. Due to the simplicity of the equation for $f_k(\eta)$, at each frozen time an unstable mode has the simple quadratic Hamiltonian of an oscillator with imaginary frequency, unbounded below, in one degree of freedom.

In principle, at each time, the Hamiltonian of the oscillatory part could be quantised in the same way as for the standard Proca system. At each wave number, gauge fixing effectively reduces the number of independent oscillatory modes to that of the usual angular momentum labels ℓ and m multiplied by three, allowing for two independent transverse oscillations plus one longitudinal oscillation. Longitudinal oscillations are allowed in an expanding space because the expansion brings an effective mass to the equivalent Proca field. However, in the unstable modes, the word "oscillation" does not apply.

For the non-oscillatory component, there is an unstable runaway. States of negative energy are unavoidable and they are not isolated by the dynamics. At each frozen time, the non-oscillatory part of the Hamiltonian has a continuous spectrum that is unbounded below. That is the same phenomenon of "jelly states" [12] in the case of M being negative and constant. The extra complication here is that $M(\eta)$ is decreasing. The level of degeneracy at each energy level n, increases as n^2. Consequently, the density of modes per unit volume, increases as n^2 [5]. The critical value n = N at the onset of instability, increases asymptotically in proportion to expansion factor $a(\eta)$. Therefore the total number of unstable modes increases asymptotically as $a(\eta)^3$ (see Ref. [2]). Assuming a sharp explosive initial condition with broad spectrum, this accounting is in accord with the energy density of dark jelly being constant. In dimensionless terms, under the exponential expansion with $a = e^{\sqrt{\Lambda/3}t}$, a wave number becomes critical when

$$k_{N,\ell} = \left(\frac{2}{3}\Lambda\right)^{1/2} a(t) \quad \sim N\pi a(t) \;\; \text{for N large.} \tag{51}$$

At the present time with $a = 1$, in terms of dimensional quantities, the length scale is R_0, which is thought to be much larger than the observable distance to the horizon of the universe. Within errors of measurement, the spatial curvature out to the horizon is zero. This uniform flatness has been explained by the inflation postulate. From the low curvature, the universe must be at least 250 times larger than the observable universe, which itself is 3.3 times larger than the Hubble radius ct_H. It is a coincidence that the Hubble time of 14 billion years is close to the age of the universe at 13.7 billion years. Through exponential expansion, t_H will remain constant as the age increases. Now the critical wave number has

$$N \approx \frac{\sqrt{2}R_0}{\pi c t_H} > 380. \tag{52}$$

Assuming that all of the independent unstable modes have evolved from independent stable modes, the number of independent unstable modes must be greater than $3 \times 380^3 > 1.6 \times 10^8$.

5. Conclusion

We have developed a generally covariant vector equation for the long-term evolution of the electromagnetic field. Such a general covariant evolution is governed by a Lagrangian density that has coupling with the Ricci curvature scalar as well as with the Ricci tensor. However our analysis includes the important case of minimal coupling in which the curvature is represented only in the Laplace-Beltrami operator that generates the flow, with both external coupling constants set to zero. Within that scheme, each choice of values for the two coupling constants defines one model. For an important class of models, the vector equation may be transformed, by a change of variables to a Proca equation with time dependent decreasing mass. With effective non-zero mass, we need to accept that universal stretching allows for the possibility of longitudinal oscillations, at least over a very long time scale.

In the long term, continued exponential expansion of the universe will surely destabilise electromagnetic fields. Successively at shorter material wavelengths, each independent dynamical mode will become repulsive at some discrete time. Beyond the present time, any dynamical mode will have only a finite number of oscillations. At any time there will be only a finite but increasing number of unstable modes, governed by a Hamiltonian in finite degrees of freedom. A dynamical mode becomes unstable only after its wavelength has been stretched by cosmological expansion, to a length that is comparable to the Hubble radius. These modes will show little spatial variation over a scale that is enormously large compared to our field of perception. However the observable universe is actually small compared to the radius of the universe beyond the horizon, that is needed in the inflationary models to explain the uniformly high degree of spatial flatness that is observed as far back into the distant past as we can see. That means that even by the current time, more than 160 billion independent modes must have become unstable. They would have little or no discernible spatial structure. Their energy would be diffuse over a very long scale, with negligible energy contained in a region that is small enough to react with laboratory-scale equipment.

Although this work is by no means complete, we may infer some properties of the quantised model from established results on quantum field theory of the Proca equation with constant mass. In keeping with the wave-particle duality, the absence of wavelike behaviour is equivalent to the absence of particle-like behaviour. At any particular time the unstable modes are governed by a quantum Hamiltonian that does not commute with any number operator. We cannot count particles in eigenstates of the time-frozen Hamiltonian. There is no state of lowest energy and there is no normalisable state that has the usual invariance properties of a vacuum. Unlike in the Gupta-Bleuler approach to the electromagnetic field, we cannot define a separate physical subspace from which the dynamics do not transgress to the unphysical sector that requires states to have negative energy and negative norm. In our experience, these phenomena of unstable quantum systems are not widely studied and not widely understood. Quantum many-body theory has been informed by knowledge of stable oscillatory systems. Fortunately the unstable component of a field coupled to expanding space-time, has only finite degrees of freedom. In order to better understand some of these properties of unstable quantum systems, Section 2 has drawn together some different approaches to the theory of quadratic Hamiltonians in quantum mechanics. The group of Bogoliubov transformations is equivalent to the symplectic group of linear canonical transformations on phase space. The angular frequencies of a linear system are invariant under those transformation groups. Therefore it is easily seen that an unstable system with complex frequencies cannot be transformed to a stable system for which the Hamiltonian is a linear combination of number operators plus a constant vacuum energy. Therefore the quantum Hamiltonian of an irreducible unstable system cannot commute with a number operator. The particle interpretation breaks down. The quantum Hamiltonian has continuous spectrum, unbounded below.

For the unstable component of the field with finite degrees of freedom, the Stone-von Neumann theorem tells us that any irreducible representation of the quantum algebra of observables as self-adjoint operators, must be unitarily equivalent to any other. The choice of representation of that component will make no practical difference. However for the stable part of the field, which has infinite degrees of freedom, the Fock representation is still available, out of an uncountable infinity of inequivalent representations.

Acknowledgements

PB is grateful to the Australian Research Council for funding of Discovery Project DP220101680.

Author details

Philip Broadbridge*†, Sarah Becirevic† and David Hoxley†
La Trobe University, Melbourne, Australia

*Address all correspondence to: author1@inbox.com

† These authors contributed equally.

References

[1] Pauli W. Relativistic field theories of elementary particles. Reviews Modern Physics. 1941;**13**:203-232

[2] Broadbridge P, Deutscher K. Solution of non-autonomous Schrödinger equation for quantized de sitter Klein-Gordon oscillator modes undergoing attraction-repulsion transition. Symmetry. 2020;**12**(6):943. DOI: 10.3390/sym12060943

[3] Bunch TS, Davies P. Quantum field theory In De Sitter space: Renormalization by point splitting. Proceedings of the Royal Society London A. 1978;**360**(1700):117-134. DOI: 10.1098/rspa.1978.0060

[4] Unruh WG. Notes on black-hole evaporation. Physical Review D. 1976; **14**(4):870-892

[5] Kodama H. On the particle-defining modes for a free neutral scalar field in spatially homogeneous and isotropic universes. Progress in Theoretical Physics. 1981;**65**(2):507-524

[6] Broadbridge P. Classical and Quantum Quadratic Hamiltonians. Adelaide: University of Adelaide; 1983

[7] Born M. Reciprocity theory of elementary particles. Reviews Modern Physics. 1949;**21**(3):463-473. DOI: 10.1103/RevModPhys.21.463

[8] Whittaker ET. Analytical Dynamics of Particles and Rigid Bodies. 4th ed. London: Cambridge University Press; 1959

[9] Williamson J. On the algebraic problem concerning the normal forms of linear dynamical systems. American Journal of Mathematics. 1936;**58**:141-163

[10] Chaiken JM. Number operators for representations of the canonical commutation relations. Communications Mathematical Physics. 1968;**8**:164-184

[11] Cook J. The mathematics of second quantization. Transactions of the American Mathematical Society. 1953; **74**:222-245

[12] Schroer B. Quantization of $m^2 < 0$ field equations. Physical Review D. 1971; **3**:1764-1770

[13] Källén G. Quantum Electrodynamics. New York: Springer-Verlag; 1972

[14] Broadbridge P. Algebraic quantization with indefinite metric. Journal of Physics A: Mathematical and General. 1983;**16**:3271-3290

[15] Friedrichs KO. Mathematical Aspects of the Quantum Theory of Fields. New York: Interscience; 1953

[16] Berezin FA. The Method of Second Quantization. New York: Academic Press; 1966

[17] Segal IE. Mathematical Problems of Relativistic Physics. Providence RI: American Mathematical Society; 1963

[18] Van Hove L. Les Difficultés de Divergences pour un Modelle Particulier de Champ Quantifié. Physica. 1952;**18**: 145-152

[19] Li L-X. Electrodynamics on cosmological scales. General Relativity and Gravitation. 2016;**48**(3):28. DOI: 10.1007/s10714-016-2028-3

[20] Francisco Cabral F, Lobo FSN. Electrodynamics and spacetime

geometry: Foundations. Foundations of Physics. 2017;**47**:208-228. DOI: 10.1007/s10701-016-0051-6

[21] Zamani F, Mostafazadeh A. Quantum mechanics of Proca fields. Journal of Mathematical Physics. 2009; **50**:052302. DOI: DOI.org/10.1063/1.3116164

[22] Birrell ND, Davies PCW. Quantum Fields in Curved Space. Cambridge: Cambridge University Press; 1982

[23] Mukhanov VF, Winitzki S. Quantum Effects in Gravity. Cambridge: Cambridge University Press; 2007

[24] Weinberg S. Cosmology. London: Oxford University Press; 2008

[25] Ruegg H, Ruiz-Altaba M. The Stueckelberg Field. International Journal of Modern Physics A. 2004;**19**(20): 3265-3347. DOI: 10.1142/S0217751X04019755

[26] Dirac PAM. Generalized Hamiltonian dynamics. Canadian Journal of Maths. 1950;**2**:129-148. DOI: 10.4153/CJM-1950-012-1

[27] Grundling H, Hurst CA. The quantum theory of second class constraints: Kinematics. Communications in Mathematical Physics. 1989;**122**:527-529. DOI: 10.1007/BF01238438

[28] Stueckelberg ECG. Die Wechselwirkungs Kraefte in der Elektrodynamik und in der Feldtheorie der Kernkraefte (III) [The interaction forces in electrodynamics and in the field theory of nuclear forces (III)]. Helvetica Physica Acta. 1938;**1**:312-328

[29] Strocchi F. Local and covariant gauge quantum field theories: Cluster property, superselection rules, and the infrared problem. Physical Review D. 1978;**17**:2010. DOI: 10.1103/PhysRevD.17.2010

[30] Delbourgo R, Twisk S, Thompson G. Massive Yang-Mills theory: Renormalizability versus unitarity. International Journal of Module Physics A. 1988;**3**(2):435-449. DOI: 10.1142/S0217751X88000163

[31] Consoli M, Pluchino A. Preferred reference system, and dragging of light in the earth frame universe. The CMB. 2021;7(8):311. DOI: 10.3390/universe7080311